ESSENTIAL

genetics

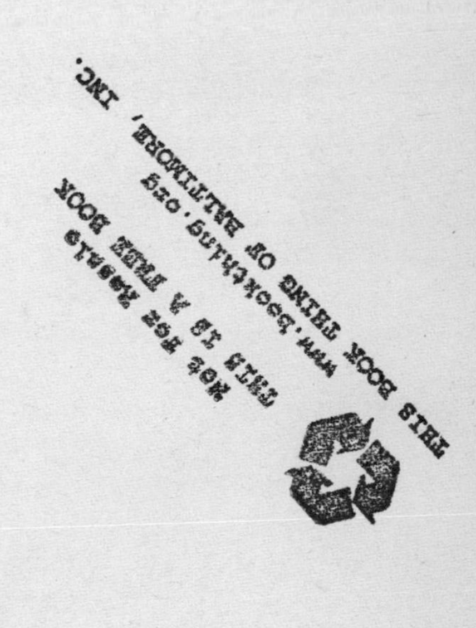

E S S E N T I A L

genetics

anna hodson

B L O O M S B U R Y

For Catto

This edition published in 1992
by Bloomsbury Publishing Limited
2 Soho Square, London W1V 5DE

A copy of the CIP entry for this book is available from the British Library.

ISBN 07475 0907 7

10 9 8 7 6 5 4 3 2 1

Typeset by Florencetype Ltd, Kewstoke, Avon
Printed in England by Clays Ltd, St Ives Plc

Contents

Introduction To Genetics

1 AN OUTLINE HISTORY

We now assume that heredity – genetics – is a part of scientific knowledge, and, this being so, that its experts are biologists and doctors. However, that assumption is new. For the greater part of human history, inheritance in the biological sense has been thought of as nothing separate from inheritance in the social or legal sense. The mysteries of why babies look like their parents, or do not look like their parents, or why some are born malformed, were in the sphere of competence (and influence) of religion and mythology.

The early Greek philosophers devised theories about heredity as they did about everything else in 'nature', which for them included the physical as well as the biological sciences. Most of their ideas about inheritance were wide of the mark, the basic mechanisms of heredity being far simpler than anyone could have guessed, given the visible complexity of the living world, but at least they were ideas. After their time, during the 2000 years of the Dark Ages and the Middle Ages, scientific endeavour was brought virtually to a standstill by the dead weight of the Church's dominance of intellectual life. In thinking about inheritance, both scientists (such as they were) and non-educated people of all classes could only go by what was presented to them by mythology, 'common sense' and the legacy of Greek philosophy. This unholy mixture unfortunately contained seven devastating misunderstandings which blocked any progress towards the truth. At least three (and probably four) of these misunderstandings are still alive today in the West and doing as much harm as ever.

Misunderstanding No. 1: The father is more important in heredity than the mother

This idea, although presumably very ancient, was first formally stated by >Aristotle. He was a man of very wide learning; his *Historia animalium* (c. 335 BC), which contains a great deal of interesting and useful biological information, was an immense advance on the state of knowledge in his day. Unfortunately, in the matter of how reproduction and heredity work, he had to rely on speculation rather than observation (an example of how different things were in biology before microscopes). His theoretical account was basically that the father provides the 'pattern' for the unborn child and the mother provides only the sustenance for it, both in the womb and afterwards. This must have had an appeal to common sense, as the process appears to be analogous to what happens with plants: you plant a seed in the ground and it grows up to look like the plant you got it from. The idea certainly also appealed to the patriarchal frame of mind of Western civilization: it stood to reason that the father, provider of status, wealth and so on was also the provider of the image of his children. (Note that this theory does not rule out the importance of a man choosing a 'good' wife; clearly, seeds planted in good soil do better than those in bad soil.) Indirectly, though of course immensely influentially, the Christian belief in the conception of Jesus has fostered this idea of image-from-father/sustenance-from-mother.

One of the great insights that >Mendel gave the world was to show that it is not true that the paternal contribution is greater than the maternal one. But the belief lingers on – and not only among racehorse breeders.

Misunderstanding No. 2: The homunculus theory

Leading on from this idea of the father as image-maker was the belief in >preformation. This is the notion that what the father implants in the mother is a perfectly formed human being, invisibly small but complete in all its parts. This idea, a very ancient one, was brought into Aristotle's formal theory of reproduction. In the case of humans, the preformed being was called a *homunculus* (Latin for 'tiny man'). No one, of course, could claim to have seen such a thing, but its existence was perhaps confirmed by the evidence that resulted from early spontaneous abortions (i.e. miscarriages). And no one ever did see homunculi, though by a strange irony one of the first pieces of biological observation that came from the introduction of microscopes did in fact reinforce the belief in them. The pioneer microscopist van> Leeuwenhoek looked at semen and saw that it was not simply a fluid but that it contained sperm. He saw enough detail to make out that each sperm has a 'tail' and a 'head', and he believed that the 'head' contained the homunculus.

It is not surprising that van Leeuwenhoek believed in preformation, as it was accepted as true throughout medieval and early modern Europe by scientists and lay people alike. It fitted in equally well with the old-fashioned patriarchal/Aristotelian idea of paternal pre-eminence as it later did with the idea of 'progress' that swept through Europe in the eighteenth century (*see* Misunderstanding No. 7 below and Acquired characteristics, inheritance of). Van Leeuwenhoek hoped that more powerful microscopes would soon be made so that a homunculus could be seen and marvelled at. The question was indeed settled by microscopes, but not within van Leeuwenhoek's lifetime, and not in the direction he had predicted. We now know that there are no homunculi, and that what is contained in the sperm (as in the female egg) is just a single set of chromosomes. None the less, maybe the old idea has not quite disappeared. Do people who say, 'Ah, he's got his father's nose', perhaps believe in some way in the actual transmission of an actual miniature nose?

Misunderstanding No. 3: The mother is responsible for the sex of the baby

An early objection to Aristotle's theory of father-as-image-former, and one which was put to him in his lifetime, was that it does not explain why many babies, unlike their fathers, are born female. His answer was that all babies would be exactly like their fathers unless 'interfered with' in some way in the womb. This interference ranged from small matters such as the baby having curly hair instead of straight to very serious ones such as it being born female or deformed. Thus the onus is put on the woman to be a 'good' mother – that is, to have sons.

This misunderstanding has, throughout history, caused many a wife to be unjustly put aside for her failure to bear sons. It has also led to a vast amount of folklore and superstition about what the mother should do to have the 'right' kind of baby, with various instructions on what she should or should not eat, should or should not look at, as well as suggestions about the timing of and positions adopted during sexual intercourse. It is also true that, at various times and places, there have been superstitions about what a man should do to beget a son (often these include orders of sexual abstinence). These superstitions could just possibly be described as slightly more enlightened, in that they admit a father's involvement in the sex of the baby, but they are not really so, for they still offer crude environmentalism for what is actually a genetic event. For the real explanation, not known to science until well into the twentieth century, *see* Sex determination. But note how often you will still hear people claiming that the number of boys or girls in someone's family is not due to chance: 'Boys run in my side of the family. My brothers and I haven't a daughter among the lot of us', and so on.

Misunderstanding No. 4: Mutants as the wrath of God

It was obvious to all observers of nature from the beginning of time that 'like begets like'. This is what happens time after time – until suddenly it doesn't. Very occasionally, a normal ram and a normal ewe, say, will produce a deformed lamb, utterly unlike either of its parents, and possibly the deformed creature is grossly abnormal, with two heads, or no head at all. Without a scientific theory, no explanation could be provided except a superstitious one, and very many societies have regarded the birth of deformed creatures as being the result of supernatural intervention: either by a god (or God) meting out punishment, or by a devil (or the Devil) getting up to mischief, or by a human practising witchcraft. Sometimes such a birth would be seen as a portent, foretelling some disaster about to fall on the family/town/country concerned. There have also been beliefs that certain kinds of abnormalities are a sign of special favour of the gods; for example, albinos are held in great esteem for this reason in some societies.

The whole idea of divine retribution being responsible for deformities is now well known to be without foundation, but none the less parents of a child who has been born with a genetic disability still very often feel shame, no matter how clearly doctors and counsellors may explain that possessing any particular gene cannot possibly be a person's 'fault'.

Misunderstanding No. 5: Spontaneous generation

At last, a misunderstanding that really has died out. In its time, however, the idea of >spontaneous generation was as big an obstacle to the understanding of heredity as any other. The idea was simply that some creatures, particularly the small and unpleasant ones, did not reproduce by breeding but arose spontaneously from their environment: fleas from dust, slugs from slime and so on. This was another theory first written down by Aristotle, who classified all the then-known creatures according to whether they bore live young, laid eggs or arose by spontaneous generation. This again must have been easily supported by common sense: in surroundings without drains or dustbins there was a great deal more filth around than we can easily imagine, and also far more rodents, bugs and pests of every kind. To say that one arose from the other was merely to put two and two together. As recently as the nineteenth century, a perfectly respectable scientist published a recipe for making mice: you put a lump of cheese in a large box with some dirty linen, and shut the lid. The problem from the point of view of the unborn science of genetics was that the idea of spontaneous generation was incompatible with a universal mechanism of inheritance: if mice spring from dirty linen, they cannot be said to inherit anything from their 'parent'.

It took a long time for spontaneous generation to be disproved; the evidence of the microscope in the seventeenth century went some way (van> Leeuwenhoek studied fleas under the microscope and saw that they reproduced sexually just as larger creatures did), but the clinching experiments were those of Louis > Pasteur in the second half of the nineteenth century. There is still a twist in the tail, however: although we know perfectly well that every living thing must have sprung from a previous living thing – how did life itself begin?

Misunderstanding No. 6: Inheritance as blending

As noted previously, Aristotle held that the father makes the most important contribution to heredity, by giving the baby its shape. Another Greek thinker had a very different theory of reproduction. He was >Hippocrates, the same person whose writings were the foundation of medicine, and after whom the doctors' oath is named. He believed that both the male and the female produce a seminal fluid (he could see it in the case of the male, and deduced that it must exist in females), which is made up of minute drops from each part of the body. At copulation, the fluids from both parents are mixed together, and the characteristics of the baby are determined by which parent's fluid prevails. This was not an all-or-nothing process, Hippocrates believed; a baby might have its mother's hands and its father's hair, or hands that were fairly like its mother's and hair of a colour half-way between that of father and mother.

This theory was much more in agreement with what can be observed in everyday life than was Aristotle's. It also provided a better explanation for the two different sexes, although those holding to the Hippocratic fluid theory should have expected that a significant number of babies would be born that were part-way between male and female. (Actually, some are born in such a condition, and Hippocrates, as an experienced doctor, was probably aware of this.) Hippocrates' theory was wrong (for the true explanation, *see* Mendel's laws), but it was a great improvement on Aristotle's, in that it stated (correctly) that both parents make an equal contribution to heredity. Although Hippocrates' theoretical and practical works on medicine had enormous and lasting influence, unfortunately his views did not prevail against Aristotle's in this case, among the scientific community, at least; among ordinary people the idea of some sort of blending seems to have taken root.

Hippocrates had believed that the hereditary elements are contained in the seminal fluid. In later centuries, people began to speak of 'blood'. This was to cause a great deal of confusion: important though blood is in so many functions of the body, it is not part of the process of inheritance. Perhaps it got into the vocabulary of inheritance as an extension of its role as one of the four 'humours' of medieval medicine; these were blood, phlegm, choler (yellow bile) and melancholy (black bile). Blood, being perhaps the most positive of these, would have been the best candidate for the fluid of generation (and the organ mainly associated with it, the heart, certainly had the right connotations). The idea of 'blood' is still current in animal-breeding circles (if nowhere else), and the word itself is amazingly tenacious: most, if not all, of the European languages still use it for expressions such as 'blue blood' (e.g. the French *sang bleu*), and even in the languages of scientists who have most definitely abandoned the concept of hereditary blood, the word itself remains in use, for instance in the term >consanguinity (from the Latin *cum*, 'with', and *sanguis*, 'blood').

Misunderstanding No. 7: The inheritance of acquired characteristics

This is the most tenacious misunderstanding of them all, being still widely believed by the general public even though a hundred years have gone by since it was proved to be impossible. The popular belief is that if a living thing – animal, plant, human, insect, what you will – acquires some particular characteristic during its lifetime, this will be passed on to its offspring. A classic example that used to be put forward was that of a proto-duck which somehow managed to spread its toes wider apart than other ducks, so that it acquired a bigger area of webbing and so swam better. This bigger web was passed on to the ducklings, who themselves strove to stretch their webbing, etc. This 'stood to reason' to early naturalists, from Aristotle onwards, and when >evolution began to be discussed, it was thought obvious that the inheritance of acquired characteristics must be the means by which evolution happens. How else could ducks have got their webbed feet, except as described above? In the case of humans, the idea also appealed to *amour propre*: if a man trains for hours every day so that he can run a marathon in under three hours, he expects his children to be born athletes. In every society, part of the point of self-improvement is to pass on the benefits to one's descendants. Unfortunately, biology is not like real estate.

However, for many millennia common sense and wishful thinking combined to support the belief in the inheritance of acquired characteristics, and at an important point in the history of science, the ideological background very strongly favoured this same idea. In the Age of Enlightenment, the eighteenth-century surge of rationalist thought in all areas, the predominant theme was that of 'progress': better societies, better machines, better roads, better sheep, better everything, to be achieved simply by applying rational principles. Political thinkers believed that human nature itself could be improved, too, and if people were to struggle in their own lifetimes to be stronger, wiser and more just, they expected that these qualities would be passed on to the next generation, and would eventually spread throughout the world. Alas: even if humankind is capable of achieving perfection (and that is a matter of faith not science), it will not be able to rely on the inheritance of acquired characteristics to carry it there. (But this doctrine was held on to for a long time in the Soviet Union – *see* Lysenko).

The reason why there is no such inheritance is discussed more fully in the entry on inheritance of >acquired characteristics, but briefly it is this: the only way in which genetic inheritance is passed from the parents to their offspring is via the sperm (from the father) or the egg (from the mother). Both of these contain essentially nothing but a single set of >chromosomes, the 'strings of beads' upon which the genes are arranged. And these chromosomes are *exact* copies of (half of) the ones that the father or mother themselves had at the moment when they were conceived. Nothing can happen to these chromosomes during an individual's lifetime in normal circumstances (for exceptions, *see* mutagen), and by the same token, there is nothing that individuals can do to improve the genetic endowment that they pass to their children. The children of the man in the example above who trained for the marathon have no better chance of being athletically gifted than if their father had sat in an armchair all day every day watching television.

Note, however, that cultural evolution (*see* exogenetic inheritance) is another kettle of fish, and a most powerful influence in humans and other social animals. In societies where the ability to swim is an essential part of everyday life, infants appear to be able to do it naturally, whereas Western children only learn with difficulty, if at all. This is a cultural, not a genetic, difference. *See also* nature versus nurture, for the more complex matter of the inheritance of abilities in advanced societies.

Another thing to notice about the theory of the inheritance of acquired characteristics is that only the inheritance of favourable characteristics was predicted. It was thought obvious that a disadvantageous acquired characteristic, such as the loss of a limb, would not be heritable. But this is to assume that the body's organs of reproduction have some sense of what is 'good' for it. Strange enough, but stranger still is the subsequent idea that the members of a species 'know' the direction which evolution of their species should take; this is known as >teleology, and is almost an important enough misunderstanding to merit a whole paragraph to itself.

Evolution as mythology

Darwin is one of the greatest figures in the history of science, and the development of genetics could not have taken place without his contribution. But he himself knew nothing about genetics, and did not study its phenomena: what he studied was evolution. The catalogue of misconceptions just listed gives an idea of the utter lack of coherence in scientific thought about inheritance up to Darwin's time (the mid nineteenth century); thought about evolution had had an equally long period of uncertainty and difficulty (*see* evolution for its modern definition).

In ancient times, there could be no theories about evolution because such a thing had not been imagined. Each species was different from all other species and each had had its own particular form since it was created. Creation myths are a fascinating study, as they contain so much spectacular imagery. These myths describe how the creatures of the living world were made once and for all just as they are now (except in a few cases in which a creature was made in one form but later changed to another as a reward or punishment). In some myths not all the creatures were created simultaneously; frequently aquatic creatures come first (the ocean is often believed to have existed before the dry land), or sometimes plants, as in the Judaeo-Christian myth given in Genesis 1:1–27, in which God makes plants on the third day, fishes, birds and whales on the fifth day, and land-dwelling creatures (man last), on the sixth day.

The earliest Greek philosophers, the Ionians, asked fundamental questions about the origins of the universe. One of them, Xenophanes, described fossil shells and fishes, and correctly concluded that they were the remains of species that had died out. That conclusion did not in itself point to the idea of evolution; it was easy to assume that, of the million (say) species created at the beginning of the world, not all are still around today, much as not all the individuals we have known are still alive. A little later, the Greek historian and naturalist Herodotus came to understand the process by which silt is deposited to form dry land, from his observations of the delta of the river Nile. That was important, because it meant that even if all creatures had been formed at once (as the Greeks believed) the land itself had not; it was still being continuously formed now as it had been in the past. The Greeks did not draw that conclusion, however, and so the two crucial pieces of the jigsaw lay side by side for 2,000 years, waiting to be fitted together.

While Christian theology had control of Western thought, there could be no speculation about the origin of the universe or of the creatures in it; the account in Genesis was considered to be unquestionably literally true. The Irish archbishop James Usher went so far as to calculate the exact time at which the events took place: the creation of man occurred on the afternoon of Friday, 28 October 4004 BC. That was published in 1650, but by then the rise of science was already threatening to undermine theological certainty.

The Danish scientist Nicolaus Steno put together the two pieces of jigsaw left by the Greeks and, in 1669, declared that *(a)* fossils were the remains of extinct marine creatures, and *(b)* that the deposits in which fossils were found had been laid down successively at the bottom of the sea in bygone ages. The theological answer to this challenge was to assert that the existence of marine fossils of extinct creatures was evidence for the literal truth of Noah's Flood. (When a giant salamander's fossilized skeleton, looking not entirely unlike that of a human being, was found near Zurich in 1726, it was named 'Man, the witness to the Deluge', because it was believed to be the remains of one of the sinful humans who were drowned while the crew of the Ark were saved.)

Then the Scottish doctor and geologist James Hutton arrived at a much more detailed understanding of how rocks are formed. He saw that the two processes of erosion and sedimentation are responsible for the formation of new layers of soil; he also believed (correctly) that ancient layers of silt become metamorphosed by the heat of the earth into new rocks, which eventually become lifted up and form new land masses. It was clear to Hutton that if all this were true, then *(a)* the formation of the world could not have been accomplished in the 5,800 years or so allowed by Archbishop Usher, and *(b)* the process of formation was continuous, 'with no vestige of a beginning, no prospect of an end'. This, the theory of >uniformitarianism, was published in 1788. It was followed in 1799 by William Smith's survey of the types of fossils found in various levels of rock in Britain; Smith concluded that fossils were always found in particular groups, and these groups were always found succeeding one another in the same order, wherever the site. Hutton had already shown that lower layers of rock were older than upper ones, and now it seemed that some kinds of fossils must be older than others. Clearly the biblical account of the creation of the physical world could no longer be acceptable to science; nor could that of the creation of the living world.

The beginning of the idea of evolution

At about this time the idea of the evolution of species was put forward, by the French scientist Comte de >Buffon, whose most important work, the 44-volume *Histoire naturelle* (1749–1804), was the first attempt at a coherent description of the whole natural world. On the evidence of vestigial organs (for instance, the toes halfway up a pig's legs), he argued that similar animals are descended from 'common ancestors'.

The idea was ahead of its time. Although progress had been made on the geological front, the biological sciences were going in a rather different direction. The advances in observation made possible by the use of microscopes had led to the accumulation of huge amounts of data, but without much theory to give it meaning. One very useful contribution was that of the Swedish botanist Carl >Linnaeus, who was interested in the problem of classification.

He began with a system of classification for flowering plants, but then went on to a global scheme of classification into which all living things could be placed: the smallest category was to be the *species*, several of which might be grouped by similarity into a *genus*; related genera went into a *class*, and related classes into *orders*. (This is still the system universally used today, but with the addition of one more level of classification – the *phylum* – in which orders are grouped together.) Thus the vast number of plants, animals, insects and other livings things known to science could be put on to the same map, as it were, and their similarities and dissimilarities discussed. Linnaeus also introduced, in 1749, the system by which each living thing could be named. Known as binomial (from the Latin, meaning 'two names') nomenclature, it consists of first a genus name which is a noun and is intended to give an idea of what 'type' of creature or plant this is, and then a name for the species (often an adjective). Thus the Linnaean system of nomenclature has *Canis lupus* for a wolf and *Canis familiaris* for a dog.

The Linnaean system was a most useful tool in preparing the way for a theory of evolution (it could of course be used for fossil forms of life as well as living ones), but Linnaeus himself was convinced that evolution could not be true. And Linnaeus was a person of influence. Even more so was Baron Cuvier (1769–1832), the French palaeontologist. Although his life's work was the collection and classification of fossil vertebrates, he rejected the idea of evolution, preferring to explain the succession of fossils by a series of catastrophes (the biblical Flood being the most recent), after each of which life was recreated (*see* catastrophism). This, though by no means in accordance with Genesis, was a great deal less offensive to the Church than the idea of evolution.

Cuvier was the most famous and authoritative biologist of his generation, but he had a formidable opponent in Jean >Lamarck (1744–1829), who was convinced that evolution was the only explanation of what fossils are. He based his argument on the fact that domesticated animals and plants are very different from the wild populations that they originally came from, and stated that a similar process must happen in nature: the environment 'requires' a new kind of animal, in the same way that humans do when they take to farming, and an existing species is gradually transformed. This process necessarily takes many generations, which explains the gradual change in fossils in the time sequence in the rocks. Lamarck was, of course, right in saying that when the environment changes, evolution occurs, and (probably) right in thinking that it happens gradually; what he was wrong about was the mechanism of inheritance underlying the process. He proposed that change in the species is produced by the inheritance of acquired characteristics (*see also* p. 9 *above*), with individuals within the old species striving to adapt themselves to the changed environment and passing on some degree of their adaptation to their offspring. The idea of the animal's will to change was part of the theory. The webbed foot of swimming birds mentioned earlier was, in fact, one of Lamarck's examples; another was the long neck of the giraffe, which Lamarck supposed to have come about as the result of pre-giraffes stretching their necks to reach higher branches.

Lamarck's championship of the idea of evolution had its effect in scientific circles, particularly as further evidence was coming in from elsewhere. Palaeontologists had noticed that in the sequence of fossils simpler life forms come earliest, and are followed by more elaborate and highly adapted ones (this became known as the principle of >succession). The late eighteenth and early nineteenth centuries were also a great time for exploration, motivated by colonialism but with a spin-off for science in that every expedition included a botanist, geologist or similar scientist. These travellers collected huge numbers of previously unknown species, and a further generalization emerged: groups of animals (or plants or insects or whatever) that live closer together are more alike than those that are far apart. For example, the scorpions that live in Africa, especially North Africa, are fairly similar to the ones in Asia but are unlike the ones in Australia or the Americas. This generalization became known as the principle of >distribution. In 1830 the British geologist Sir Charles >Lyell published *The Principles of Geology*, in which he showed conclusively that the process of rock formation was continuous, as Hutton had claimed a generation before, but whereas Hutton's ideas had remained controversial, Lyell's better exposition of them was immediately accepted.

One more factor deserves to be mentioned, and that is the stranglehold that the Church of England still had on public and intellectual life in Britain. Until 1828 no one could take public office who was not a communicant member of the Church, which excluded Jews, Roman Catholics and every sort of Nonconformist, as well as mere non-believers; and it was a further 50 years before such people were allowed to hold a teaching post at a university. (This was one reason that most good science of the time was being done by amateurs.) And if the Church had tacitly had to concede on the creation of the physical world, it was determined to take a stand on the creation of life.

Darwin

That was the background to the work of Charles >Darwin: geological evolution had become a certainty, biological evolution a near-certainty but not yet proven. Darwin was not a trained biologist, but after leaving university he went on an expedition (on HMS *Beagle*, 1831–6) as an unpaid naturalist, to make observations while the ship carried out an oceanographic survey for the Navy. He was interested in the fact that there are many varieties to be seen within a single species, and to record this phenomenon he collected thousands of specimens and made volumes of notes. He was particularly struck by what he found in the Galápagos Archipelago, a group of 15 small islands 500 miles (800 km) west of Ecuador. The plants, birds and animals there are similar in a general way to the ones found on the mainland, but every species is recognizably different from the mainland form. Further, the types are often very different from one island to the next, Darwin found that the giant tortoises (from whom the archipelago gets its name, *galàpago* being old Spanish for 'tortoise') were so distinctive that one 'could with certainty tell from which island any one was brought'. When Darwin came home, he published his *Journal*, and took his time over arranging and editing his notes.

Darwin was sure that evolution was a reality, and like Lamarck he felt that the selective breeding practised by farmers to develop breeds of animals was a clue to the process by which species changed during evolution. 'But how selection could be applied to animals living in a state of nature remained for some time a mystery to me', he wrote in 1837. Then he read the work of the eighteenth-century economist Thomas >Malthus, who postulated that, since human populations increase faster than the food supply (a pressing issue in Malthus's time, with the Industrial Revolution being associated with desperate problems in growing enough food for the suddenly much more numerous urban population), poverty and distress are inevitable. The only mechanisms that prevented the human population expanding indefinitely, in Malthus's view, were war, famine and disease.

Suddenly it all fell together in Darwin's mind. He had noticed that populations of animals and plants in the wild do not tend to increase, although each individual animal or plant that breeds produces far more offspring than are needed to replace itself. If more young are born than can survive, then there is a struggle (Malthus's war, famine and disease). However, what if success or failure in this struggle is not just a matter of chance? Darwin saw what was the meaning of the variations he had observed: 'Under these circumstances, favourable variations would tend to be preserved, and unfavourable ones to be destroyed. The result of this would be the foundation of new species.' That was written in 1838.

Darwin came to the conclusion that there was a force that he called 'natural selection'. He believed that in an unchanging environment, natural selection would favour the conforming type of animal within the species, and that divergent variations would lose out in the struggle for survival; thus the species would preserve its identity. But if a change in the environment occurred, those divergent variations that happened best to suit the new conditions would be favoured by natural selection, and the former type would lose out; thus the old species would change gradually into a new one.

Darwin thought long and deeply about this, and came to the conclusion that it was the only explanation that fitted with the succession of fossils through geological time and with the geographically dispersed variations of species that he had seen on his travels.

Darwin realized that what he had discovered was completely counter to the teaching of the Church, still holding the line on the Creation, and that to publish his views would cause a furore. He also saw clearly that human beings were not a special case in nature; if every other species had evolved from a more primitive ancestor, so had man. There had been no Adam: there must have been an ape. Darwin was extremely worried by this. Though he was convinced of the truth of his theory; he shrank from the thought of proclaiming it; he was a shy and conventional man, and a semi-invalid, and he had no wish for himself and his family to be in the centre of a shocking controversy. He consulted his friends (who included some of the leading scientists of the day, men like the geologist Sir Charles Lyell and the botanist Sir Joseph Hooker), who urged him to publish. Darwin still hesitated, however, and turned to other work.

In 1858, *twenty years* after he had arrived at his theory, something unexpected happened. Darwin received a manuscript from a fellow naturalist, Alfred Russel >Wallace, who was at the time collecting specimens in Malaya. Wallace had hit upon exactly the same theory of evolution by natural selection (and his draft paper on it was considerably more succinct than the treatise that Darwin, reluctant to finish, was still only half-way through writing). Darwin's friends arranged for Wallace's paper and an abstract of Darwin's treatise to be presented together at a meeting of the Linnaean Society in London on 1 July 1858. Wallace and Darwin were on good terms and did not enter into a dispute about who had the right to claim the theory as his own. Wallace, much younger and from a lower social class, was quite happy to stand back and let the well-connected Darwin attract the undesirable attention that was bound to follow.

So in the following year Darwin published the full account of the theory in the book whose full title is *On the Origin of Species by Means of Natural Selection, or the Preservation of Favoured Races in the Struggle for Life*. There was indeed an uproar, not only from the Church but also from all parts of the Establishment, up to and including the Cabinet (William Gladstone, then Chancellor of the

Exchequer, was a bitter and vociferous enemy of Darwinism). In the face of skilled and committed defence of Darwin's theory by T. H. >Huxley, the religious opposition was reduced to the dangerous tactic of ridicule, most notably when Bishop Wilberforce attended the meeting of the British Association in Oxford in 1860, thinking that his trump card would be to ask Huxley whether it was through his paternal or maternal grandparents that he was descended from a monkey. It is hard to imagine, with the absolute supremacy and authority of practising scientists that is taken for granted in the late twentieth century, that the Church should still have been able to intervene in the theoretical side of science in this way; the fact that it is no longer able to do so is almost entirely due to the great struggle over the theory of evolution. (The Church, and religions of all sorts, still have an important and legitimate voice in the applications of science.)

There were, however, two points in his theory for which Darwin realized that he did not have a satisfactory explanation. Firstly, he did not understand how variant forms arose. He did not believe, as Lamarck had done, that they were 'called forth' by the environment itself; his observations had shown him that the variants are entirely random with respect to the environment (this is understood now: *see* mutation). Secondly, he realized that his theory required a mechanism of inheritance that was stable. Lamarck's theory of the inheritance of acquired characteristics was not a possibility for him, as he did not believe that the characteristics in question were acquired; for him they were innate. Darwin therefore sought some theory of inheritance other than the widely accepted Hippocratic 'blending', which would not do because, if it were true, the advantage of belonging to a new variant form would be 'diluted' at every generation. This lack of a suitable mechanism of inheritance was to worry Darwin to the end of his life.

Mendel

In fact, the mechanism of inheritance had already been discovered, in 1866, not long after Darwin's publication of *The Origin of Species*. The discoverer was an Augustinian monk named Gregor >Mendel, living in what is now Czechoslovakia. Darwin was not the only person in the scientific world who did not hear of Mendel's work: the crucial paper made a modest appearance in the

Proceedings of the Natural History Society of Brno and did not become known outside the small circle of that society's members for the next 35 years.

Mendel was not a sheltered amateur pottering in the monastery garden; he was a trained scientist and taught mathematics at a local school. He did not believe that inheritance could possibly be a matter of blending, since he had observed in his plant-breeding experiments that in hybrid strains of plants the original parental types often re-emerge unchanged in later generations; this would not be possible if the original types had been blended to form the hybrid. It is true that many other people had noticed this too (the word 'throwback', used to describe this phenomenon, was established in English in the nineteenth century before Mendel's work was known), but Mendel was the first to devise a theory to explain it.

He postulated that inheritance is under the control of independent particles, which he called 'factors' (we now call them 'genes'). For every characteristic – for example, flower colour, stem length, pod shape, to use examples from Mendel's peas – the individual possesses two genes, one inherited from each parent. When the individual comes to make its gametes (an extremely convenient term denoting the sperm or ovum in animals and humans, or the ovule or pollen grain in plants, i.e. the semi-cell in which the individual transmits its genes to the next generation), it places *one* gene for each characteristic in each gamete. Thus when two gametes unite at fertilization, the resulting embryo has again *two* genes for every characteristic; the two genes not only do not blend, they do not affect each other by co-existing in this individual. It seems that Mendel thought of this utterly simple scheme in the abstract, rather than coming to it as the result of experimentation. But one more feature of his theoretical scheme enabled him to put it to experimental test: he postulated that when the individual makes gametes it makes *equal quantities* of gametes having the maternal or paternal gene for each characteristic.

On this basis Mendel was able to predict what would happen in a cross between two pure strains of peas. Taking a simple characteristic such as the shape of the seed, he predicted that if he were to take two pure strains, one with round seeds (genes *RR*) and one with wrinkled seeds (genes *rr*), all the first-generation (or >F_1, for first filial generation) hybrid plants would have genes *Rr*. These plants had round seeds, identical in appearance to those of one of the parental pure strains. Mendel described this situation as being due to one gene being >dominant over the other, the latter being >recessive. (These are the terms still in use today, and Mendel's notation of capital letters for dominant and lower-case for recessive is also still used.) Mendel believed that dominance and recessivity did not in any way affect the basic mechanism by which the two genes are separated in equal quantities into the gametes. This he proved by showing that, in a cross between F_1 hybrids (all looking like the parental strain having the dominant characteristic), the >F_2 hybrids (i.e. the second generation) are in the ratio of 3:1, with three plants showing the dominant type for every one of the recessive type.

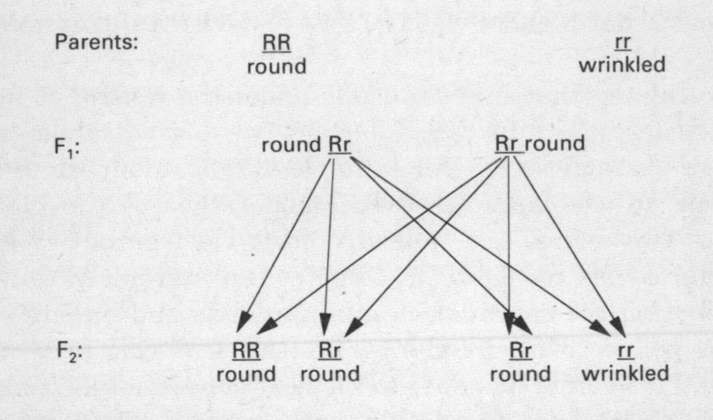

After counting the results of literally hundreds of plant crosses Mendel found that it is always true that, on average, the F_2 generation contains three times as many plants of the dominant type as the recessive. He repeated the experiment with six more characteristics, all of which gave the same result. He also proved that of the plants showing the dominant type one in three will 'breed true' (because it has inherited two of the dominant genes), while the other two resemble those of the F_1 generation in having one dominant and one recessive gene. In the bottom row of the diagram above, the individual on the left will breed true, having only dominant genes, but the two in the middle will not.

These results confirmed Mendel's basic ideas: a plant has two self-perpetuating hereditary factors (i.e. genes) for each characteristic; the plant's gametes contain only one gene per characteristic; and the 'paternal' and 'maternal' genes occur with equal frequency among the gametes. This last fact – the way in which genes are divided evenly during the formation of gametes – is known as Mendel's First Law, or the law of >segregation.

Mendel had a further insight of almost equal importance to the first great one of segregating 'factors'. It was this: the genes controlling one characteristic are inherited independently of the genes controlling another characteristic in the same plant. For example, Mendel believed (correctly) that the height of a plant and the colour of its flowers are inherited independently of each other. Some peas are tall (dominant), while others are dwarf (recessive); some have purple flowers (dominant), others have white (recessive). In a cross between two strains, each starting off pure for the two different traits, Mendel believed that the F_1 generation would all show the dominant type for both traits, but that in the F_2 generation every possible combination would be found. This would only be true if the genes were independent and did not 'stick together' by virtue of having been inherited together from the same parental strain. This is Mendel's Second Law, the law of >independent assortment.

This theory predicted that for two independently assorting characteristics, each segregating at a ratio of 3:1, the types seen in the F_2 generation would be in the ratio of 9:3:3:1. That is, there would be 9 of the type dominant for both characteristics, 3 dominant types for one characteristic but recessive for the other, 3 more likewise but the other way round, and 1 recessive type for both characteristics. How this works can be shown in a diagram called a Punnett square (for an example, *see* dihybrid). Mendel did get results confirming this prediction.

The remarkable thing is that although this theory and its statistical proof were clearly set out in Mendel's paper, and although the journal in which it appeared was subscribed to by the major libraries of Europe and the United States, the scientific world did not appreciate its significance. It is also remarkable that the same theory was discovered again not once but three times simultaneously. At the very end of the nineteenth century three botanists – de >Vries in Amsterdam, >Tschermak von Seysenegg in Ghent and >Correns in Tübingen – independently obtained the same results in their experiments as Mendel had done. Even more remarkably, each of them then searched the literature and found Mendel's 34-year-old paper. De Vries saw to it that Mendel's work was republished (in 1900) and that the credit for the theory went where it was due.

Genetics becomes a science

For Mendel, the 'factors' of inheritance had been entirely abstract entities. He did not trouble to speculate about what their physical reality would turn out to be. As it turned out, their material basis was proved within a couple of years of the rediscovery of his theory. Microscopes were by that time operating at magnifications of × 1000 or more, and the fine details of the structure of cells in plants and animals were being described and photographed. Structures in the cell nucleus were seen that were like dark-coloured threads; they were named chromosomes (from the Greek for 'coloured bodies'). Before long, in 1902, the American zoologist William Sutton noticed that in the gametes there are half the number of chromosomes found in ordinary cells, and that the set in a sperm almost exactly matches the set in an ovum. It was immediately realized that here was the reality behind Mendel's 'factors'. These chromosomes were seen in the cell nuclei of all sorts of living creatures – in insects, animals, humans, as well as in plants – so it was clear that the Mendelian machinery of inheritance was the one that operated universally.

The terminology of genetics developed so as to be able to name the processes involved in Mendelian inheritance. The alternative versions of a gene for any particular trait, or characteristic, are called >alleles. If an individual has two identical alleles for a given trait, like the wrinkled-seeded pea in the example above (*rr*), it is said to be a >homozygote; if its alleles are different (*Rr*), it is a >heterozygote. An individual's genetic make-up is described as its >genotype; this can either refer to the alleles for a single trait or it can cover two or more at once, for example, a pea plant with a dwarf habit and yellow peas has genotype *dd gg*. Related to the genotype but importantly different is the individual's >phenotype, that is, its appearance or physical characteristics. The phenotype does not have a one-to-one relationship with the genotype. Where one allele is dominant (as in all Mendel's examples) the heterozygote with one dominant allele has the same phenotype as the homozygote with two; thus pea plants with the same phenotype of a tall habit may have genotype *Dd* or *DD*.

Further microscopical observations revealed that the chromosomes are involved in two different kinds of cell division. The first happens during growth or tissue renewal, when an ordinary cell within the body divides to form two new cells exactly like itself: this is called >mitosis. In this process each of the new cells receives a set of chromosomes identical to the set in the original cell. The other kind of cell division is what happens in the formation of gametes, and is called >meiosis. In this process the chromosomes first align themselves in pairs, the chromosomes inherited from each parent lying alongside each other; then, during the division, one of each pair moves (at random) into one of the two gametes being formed. These observations provided the physical evidence for Mendel's hypothesis (his law of segregation), in which he stated that equal numbers of the two kinds of gametes are formed.

Chromosomes

Chromosome research became the focus of genetics. One early line of investigation was to discover the numbers of chromosomes in the cell nuclei of different species. These turned out to be surprisingly variable, and were not in any simple way related to complexity (or evolutionary 'advancedness'); for instance, humans were found to have 46, dogs and chickens 78, scorpions 4, peas 14, bracken 116, some moths over 200. Notice that all these are even numbers, being the sum of the maternal and paternal sets; one could equally well express these numbers in pairs, so that humans have 23 pairs, dogs and chickens 39 pairs and so on. However, early studies on the chromosome pairs revealed that not all are exact pairs: although in female animals, all the pairs match, the chromosomes in one pair in males are not exactly the same. It was soon realized that this provided the explanation for how sex is determined: whether an animal turns out to be male or female depends on which chromosomes it has in this particular pair. The sex chromosomes were designated X and Y, so that the female has XX and the male XY (the non-matching pair). When the female forms eggs, every egg receives a single X chromosome, but when the male makes sperm, half receive the X chromosome and half the Y chromosome. From this it follows that the sex of the offspring depends on whether it was produced by an X-bearing or a Y-bearing sperm; therefore it is the father and not the mother that determines the baby's sex (*see* sex determination).

Each member of a pair of chromosomes is called the homologue of the other; the X and Y chromosomes, however, are non-homologous (though the two X chromosomes in a female are homologous). The ordinary chromosomes are known as >autosomes. The complete number of chromosomes – that is the paired individuals that are found in all the cells of a living thing – is known as the >diploid number; the number making up a set consisting of only one from each pair, as found in the gametes, is the >haploid number.

Although it was obvious that the chromosomes provided the physical basis for Mendel's mechanism of inheritance, there was, one more difficulty. The chromosomes could not be the same as Mendel's 'factors' and our 'genes' for the simple reason that there were not enough of them. There are seven pairs of chromosomes in the kind of pea that Mendel studied – but the pea must have thousands of different genes governing all that plant's many characteristic. It was soon realized that the chromosomes must represent not single genes but strings of very many genes connected together.

But if this were so, how could Mendel's second law, the law of independent assortment, be true? How could the genes for two different traits assort independently if they were physically located on the same chromosome?

The answer is that they can't: if the genes are on different chromosomes, they do assort completely independently, but if they are on the same chromosome, they travel together. Mendel had simply had extraordinary luck in choosing to investigate seven traits that each happened to be on a separate chromosome; therefore he observed independent assortment. Other investigators looking at the progeny in a cross involving two traits had obtained quite different results. They saw that in many such crosses the genes for two different traits were inherited together, as a package, and did not assort independently. The reason, as they correctly deduced, was that these genes were located on the same chromosome: this is known as >linkage, and each gene's position on the chromosome is its >locus. But another phenomenon also emerged from these crossing experiments: occasionally the 'package' of alleles linked together on the same chromosome gets split up, and alleles that you would expect to find in the same gamete end up in opposite ones. This is because during the meiosis cell division the maternal and paternal chromosomes lie alongside each other and often swap sequences of alleles with each other. After this process (known as >crossing-over) the two chromosomes still have one allele at each locus and in the right order, but not the same alleles as they started out with throughout their length. This was first suggested by de Vries in 1903, but the idea was fully worked out and experimentally confirmed by the American geneticist T. H. >Morgan in 1910–11.

Mapping the chromosomes

A student of Morgan's, A. H. >Sturtevant, made the brilliant deduction that if cross-overs occur at random along the length of the chromosome, the genes lying close together will seldom be separated by a cross-over, while this will happen often to genes at opposite ends of the chromosome. In other words, the cross-over frequency can be used as a measure of distance between loci so as to make a map of a chromosome. Sturtevant made his first chromosome map in 1913; it showed the relative positions of six loci on the X chromosome of the *Drosophila* fruit fly. By 1922 Morgan, Sturtevant and their associates had mapped over 2,000 loci, on all four of *Drosophila*'s chromosomes (*see* genetic map). This all had an impact on evolutionary theory, as it was realized that crossing-over could provide one source of the genetic variation needed as the basis for Darwinian evolution. Instead of the maternal and paternal chromosomes marching on down the generations unchanged, they shuffle the alleles between them. The alleles remain the same, but in different combinations and as a result, they may suddenly produce a phenotype previously not seen.

Mutation

At the same time another, and even more important, source of genetic variation was being investigated. In 1915 another American geneticist, H. J. >Muller, was working on *Drosophila* breeding experiments similar to Morgan's when he noticed that, very occasionally, an allele emerges that is completely new and cannot have been inherited from either parent. He coined the word >mutation (from Latin, meaning 'change') for this sort of newly arisen variant allele. Muller correctly thought of a mutation as being some sort of chemical change within the organism. Observing that once a mutation has arisen, it is inherited in the ordinary way and does not change back to a normal allele, he saw that the occurrence of mutations in species in the wild was the ultimate source of genetic variation and therefore of evolution itself.

Muller then became interested to know how often mutations

occur. He realized that if they were too frequent, the stability and continuity of a species would be threatened, but that if they were not frequent enough, there would not be enough available variation for a species to adapt itself genetically in response to changes in its environment. The rate of mutation in nature is normally very slow – the genetic machinery tends towards the conservative, not the innovative. However, Muller discovered (in 1926) that in certain circumstances the mutation rate can be speeded up enormously: when he irradiated *Drosophila* flies with X-rays, he obtained very high rates of mutation (and the higher the X-ray dosage, the more mutations occured). Other types of radiation and some chemicals were found to have the same effect (*see* mutagen). Muller's work showed that the great majority of mutations result in changes that are detrimental to the organism (as would be expected, since they are random changes to a previously well-organized 'machine'), but that fortunately most mutations are recessive.

Mendelian inheritance in higher animals

While Morgan, Sturtevant, Muller and others were expanding the knowledge of genetics enormously with their studies on *Drosophila*, others were extending Mendelian genetics to other organisms. Because of the practical implications, much experimental work was done on plants that are food crops in the hope that genetics could be used to improve yields, develop disease-resistant strains and so on (such had also been Mendel's motive for undertaking his research). Thus there was much research done on maize (sweetcorn, *Zea mays*) and wheat. It turned out that many of these plants have multiple sets of chromosomes (>polyploidy), which gives an extra source of variation, while still adhering to Mendelian principles.

The first mammal in which Mendelian inheritance was studied was the mouse. Cattle and sheep are obviously the mammals of practical importance, but their generation time (that is the time from the birth of an individual to the birth of its first offspring) is much longer, which makes genetical studies no more difficult but just more time-consuming. It turned out that, although some traits such as coat colour can be traced using Mendelian principles, many of the economically significant traits, such as growth rate, are not controlled by single genes.

It came as no surprise when Mendelian genes were also proved to exist in human beings (humans had not been a special case in Darwinian evolution; nor were they in Mendelian inheritance). The first evidence for this came in 1903 from a British doctor, Sir Archibald >Garrod. He looked at the pedigrees of sufferers from certain diseases that seemed to run in families. He realized that the diseases which he called >inborn errors of metabolism could be explained as being caused by a recessive gene, with only the homozygotes suffering the illness while heterozygotes were normal. Garrod's idea was that these diseases were caused by a simple chemical 'misprint'. He further proposed that this misprint resulted in non-production of a single >enzyme: he was right, but his findings were ignored for nearly 40 years.

While Mendelian inheritance was well established as occurring in every sort of living thing, there was still the problem of the very many characteristics that did not appear to be controlled by single alleles. As mentioned above, such characteristics include most of the economically important features of animals (and many of the socially significant features of humans). These are known as >continuous variables, as opposed to the >discontinuous variables that Mendel had studied. Many geneticists were inclined to believe that continuous variables had a blending rather than a Mendelian machinery of inheritance. This was finally disproved by the British mathematician Sir Ronald >Fisher, who analysed statistical data from agricultural field trials and came to the conclusion that the data could only be explained by the proposition that continuous variables are controlled by a small number of separate genes which interact with each other and with the environment. There had been many scientists of the opinion that Mendelism was too rigid to allow evolutionary progress, and that some blending or inheritance of acquired characteristics had to be involved. However, Fisher in his masterpiece of genetical argument, *The Genetical Theory of Natural Selection* (1930), proved in mathematical terms that Mendelian inheritance is the only possible basis for Darwinian evolution.

Genes as 'messages'

Although the behaviour of genes was becoming better and better understood, there was no information about their physical behaviour beyond the fact that they were located in a row along the chromosomes. It was clear that the genes were in some way carrying 'messages', and that in order to be self-perpetuating through countless cell divisions, the genes must be able to duplicate themselves exactly and accurately. But there was no theory about what chromosomes could be made of to give them these remarkable properties. A clue came from an experiment in the 1920s with two strains of pneumococcus bacteria (which cause pneumonia). It was found that there was a genetic 'transforming factor' by means of which bacteria of a virulent strain, even after they had been killed by heat, could pass on their virulence to a non-virulent strain.

It was many years before this 'transforming factor' was isolated and identified, and when, in 1944, the American bacteriologist O. T. >Avery announced that it was deoxyribonucleic acid (>DNA) he was resoundingly ignored. This was partly because very little was known about the chemistry of DNA except that it consists mainly of phosphate, the sugar deoxyribose and four nitrogenous bases: adenine (A), cytosine (C), guanine (G) and thymine (T).

One person who followed up the clue that DNA could be the stuff that genes are made of was the American biochemist Erwin >Chargaff. He analysed DNA from many sources, and found that each species has its own characteristic type of DNA, in terms of the relative amounts of the bases A, C, G and T. Another of his findings was that within any sample of DNA there was always the same amount of A as of T, and the same amount of C as of G. This was in 1950, and at the same time the British crystallographers Rosalind >Franklin and Maurice Wilkins were taking X-ray pictures of crystallized DNA molecules showing that DNA could assume a helical or spiral structure, though it was not clear whether this helix had two or three strands.

The matter was soon settled, in Cambridge in 1953, by the British biologist Francis >Crick and the American zoologist James >Watson. Using Chargaff's findings plus Franklin's and Wilkins's X-ray crystallographic data, they tackled the problem of DNA's structure by building a scale-model made of wire and pieces of cardboard. The molecule was like a spiral staircase, with the steps made of pairs of bases (A always joined to T, C always joined to G) and the banisters made of sugar–phosphate chains. They described this in a marvellously understated paper in the scientific journal *Nature*, which contained the words: 'It has not escaped our notice that the specific pairing we have postulated immediately suggests a possible copying mechanism for the genetic material.' That paper signalled a complete change not only in genetics, but in the place of genetics in the life sciences, and the place of the life sciences in the world of knowledge.

2 AFTER CRICK AND WATSON

At the simplest level, what Crick and Watson had discovered was that DNA had a structure which allowed it to copy itself. Since A must be paired with T, and C with G, it follows that, if a DNA molecule is split down the middle lengthwise, all the information is there to reconstitute the whole molecule again (*see* replication).

Crick and Watson had in fact discovered more than this. Their proposed structure for DNA also provided the explanation for how a gene 'works', in chemical terms. It was already known that what a gene does is to control the production of a single >protein. This had been neatly shown in 1941 by the Americans E. L. >Tatum and G. W. >Beadle, who had looked at mutant forms of the bread fungus *Neurospora crassa* and had found that each single mutant resulted in the loss of the ability to make a corresponding single enzyme. (This idea – 'one gene, one enzyme' – had first been thought of by Sir Archibald Garrod nearly 40 years earlier.) What the Crick–Watson model showed was how the gene – a length of DNA – could contain a message encoded in the sequence of letters must be the specification of which protein is to be made. So, the coding structure of DNA does two things:

- It ensures that DNA is replicated to produce more DNA.
- It ensures that DNA is transcribed into RNA, which is then translated into protein.

This twofold scheme is known as the >central dogma of molecular biology, and was first proposed by Crick. On it rests the entire modern understanding of the life sciences.

The process of DNA (i.e. the genes) controlling the production of proteins happens continuously throughout the body, and is the essential basis of how the body works (in an animal, plant, insect, any living thing, even a single-celled creature). One could therefore say that what DNA is really 'for' is the running of the body's biochemical functions; when it acts as the vehicle of inheritance it is only passing on the wherewithal for the next generation to run the body's functions. Of all the DNA molecules that a human being makes in a lifetime, the vast majority (many millions to one) will be involved in making the proteins in the body's cells; only a few will go into the gametes. To make an imperfect analogy, DNA is like a small businessman, who does all the jobs around the business himself, some things on some days, some on others, but of all the working days of his life, he spends only one or two telling his son what to do when he takes over.

Transcription and translation

When DNA makes a protein, it does so via an intermediate molecule called ribonucleic acid (>RNA) which is very similar to DNA. Its sugar backbone consists of ribose, not deoxyribose, and the pairs of bases are adenine (A) and uracil (U) (instead of thymine (T)), and cytosine (C) and guanine (G). RNA is also usually in the form of a single strand, not a double-stranded helix like DNA. A molecule of RNA can be made to an exact and repeatable pattern by reading off the base sequence of a stretch of DNA; this process is called >transcription. The strand of RNA produced in this way is known as >messenger RNA, or mRNA, since it carries the message telling what protein is to be made from the particular stretch of DNA that was copied (*see* protein synthesis).

All proteins consist of a chain of sub-units called >amino acids. There are some 20 amino acids commonly occurring in protein; every protein has its own sequence of amino acids, from as few as half a dozen (which is not really a protein, but a >polypeptide) up to as many as a thousand or more. The sequence determines not only the chemical activity of the protein but also its physical shape; proteins are wound into their own characteristic balls or spirals. The sequence of amino acids in the protein is dictated by the sequence of bases in the DNA, as was immediately obvious when the Crick–Watson structure was proposed.

Much of the research on working out the details was done, as had been Crick and Watson's, at the Medical Research Council's Laboratory of Molecular Biology in Cambridge, England, which in the 1950s was a leading centre of biological research. A South African, Sydney >Brenner, worked with Crick on deciphering the DNA code, and they discovered that it is written in 'words' of three letters. For example, a sequence of DNA in which the bases are GCATAGATC is to be read as GCA–TAG–ATC (these groups of three are known as >codons, or triplets). It seemed likely that one codon coded for one amino acid, and the truth of this was first demonstrated in 1961 by the American scientist Marshall >Nirenberg who, having made a strand of synthetic mRNA consisting of repeated UUU codons, obtained a synthetic 'protein' consisting of one amino acid (phenylalanine) similarly repeated. There followed a worldwide race to decipher the other codons, with many of the results coming from the Indian–American geneticist Har Gobind >Khorana. As there are four possible bases occurring in groups of three there are 64 (i.e. 4 × 4 × 4) possible combinations, but only 20 amino acids to be coded for. It turned out that most amino acids are coded for by more than one codon, and that there are three codons which do not represent any amino acid but are 'stop' signals where the protein-coding message ends (*see* >genetic code).

Further research demonstrated that this genetic code is the same in all living things. This is the strongest possible evidence that all life on earth is descended from a single origin.

Apart from some minor differences between >prokaryotes (which include very simple organisms such as bacteria) and >eukaryotes (the more advanced creatures, including humans, that have a proper nucleus in their cells), the way in which proteins are made is the same in all living things. The code sequence of the DNA gene is used to build a molecule of messenger RNA; this is assembled by an enzyme called >RNA polymerase, and this part of the process is called transcription.

There are also many molecules of a different kind of RNA known as >transfer RNA, or tRNA. Each tRNA molecule consists of only three bases, that is, it is a single triplet. These three bases form an >anticodon, and each of these matches on to a codon in the mRNA. Thus if the mRNA reads GCA–AAG–AAA, it will attract to itself three tRNA molecules having the anticodons CGU–UUC–UUU, and each of these has the appropriate amino acid attached to its other end. This part of the process, known as >translation, takes place on one of the many lumps of protein-and-RNA in the cell called >ribosomes, which act as a sort of assembly frame for building proteins. The tRNA molecules form a line, and the amino acids join up in the specified order to form the protein chain.

Once this machinery was understood, both mutation and dominance became explicable in molecular terms. If just one base in the DNA changes (in a mutation), the codon also changes and is very likely to code for a different amino acid. The difference that this new amino acid, being only one out of so many, makes to the resulting protein is very small but it will probably have a bad effect on its function, even to the point where the protein does not work at all. (Very occasionally, such a change will make the protein more efficient than it was before.) Sometimes the mutation alters the intended codon to a 'stop' signal, in which case the formation of the protein ends before it is complete. There are also >frameshift mutations, in which the process of transcription starts in the middle of a codon rather than at the beginning, so that the entire message is garbled and a useless protein results.

Equally, it became clear why most mutations are recessive. When a mutation occurs on one chromosome of a pair, which then starts coding for a defective protein, it does not matter so long as the gene on the other chromosome can produce adequate quantities of the normal protein. This is normally the case if the protein in question is an enzyme, which is only needed in small amounts, but for structural proteins, such as the ones that form bone tissue, the output of a single gene will probably not be enough. Thus the enzyme-defect mutations resulting in diseases such as >phenylketonuria are recessive, while structural-protein mutations such as those causing >achondroplasia are dominant.

Gene regulation

Even after the statement of the central dogma of molecular biology, there was still a major difficulty in understanding how genes work. Every cell in the body has the same set of genes, located on the same paired set of chromosomes that have been copied exactly in every cell division since fertilization. Yet cells are specialized: some are skin cells, some line the intestines, some are nerve cells and so on. It must be that genes can be switched off when they are not needed (and probably most genes, about 90 per cent, are switched off most of the time). How does the switch work?

This was first investigated in the common bacterium >*Escherichia coli*, which has a simple metabolism but needs different enzymes at different times according to what nutrients it has to digest. In 1961 two French molecular biologists, Jacques >Monod and François >Jacob, discovered that the switch mechanism consists of what they called an >operon: a group of genes with related functions have an 'operator' sequence next to them; this is switched off by a 'repressor' molecule which physically combines with the operator, thereby preventing mRNA being formed to transcribe the genes. Repressors (like genes themselves 60 years before) remained only hypothetical entities until in 1968 the American molecular biologists Walter >Gilbert and H. Muller-Hill successfully isolated a repressor and showed it to be a protein.

Operons and repressors are now well known in prokaryotes, but they have not been definitely shown to work in the same way, or indeed to exist, in eukaryotes. Prokaryotes have another way of switching genes on and off, called >attenuation, but this also has not been proved to exist in eukaryotes. In some insects (including *Drosophila*), all of which are eukaryotes, the chromosomes in the cells of the salivary glands are so large that they can easily be observed, and in the areas where DNA is being transcribed they can be seen to swell up into a >puff. Although these puffs are known to be the site of gene activity, that does not explain how these genes are switched on and off.

One possibility is that gene regulation in eukaryotes is connected with the phenomenon of >jumping genes. Discovered in maize plants in the 1940s by the American geneticist Barbara >McClintock, jumping genes are segments of DNA within a chromosome which have flipped over so as to alter their position in relation to neighbouring genes on the same chromosome. McClintock proposed that this might enable the genes that 'jumped' to act as switches for the neighbouring 'working' genes. Her idea was ignored at the time, but later research in the 1970s showed that jumping genes, now known as >transposons, are common in both prokaryotes and eukaryotes, and it is possible that they are indeed involved in the regulation of gene activity.

Movable genes and cancer

It may also be that movement of genetic elements within the chromosomes is a factor in cancer. The clue came from the fact that some cancers are caused by viruses; this was first discovered in chickens, who develop tumours when infected with the Rous sarcoma virus. The virus introduces a gene – known as an >oncogene – into the chicken cell, which transforms it into a cancer cell. It was then discovered that all normal cells contain many genes that are exactly comparable to viral oncogenes; these are known as proto-oncogenes, and they seem to play a part in the control of cell division and differentiation. There is evidence to suggest that different types of cancer are caused by the activation of different proto-oncogenes; this activation might in turn be caused by a viral oncogene or by a mutation of the cell's own proto-oncogene on contact with a harmful substance.

For example, Burkitt's lymphoma, a cancer causing tumours of the jaw, which is common in Africa, occurs in children who have been infected with the Epstein–Barr virus; the tumour cells have a chromosome abnormality, with a translocation broken off from chromosome 8 at exactly the point where the proto-oncogene analogous to the virus is situated. It is plausible that the virus enters a single cell and interacts with the proto-oncogene in such a way as to cause chromosome breakage and translocation; this moves the proto-oncogene into a new position where it is not under its normal control mechanism, so the cell in which this occurs proliferates and gives rise to a mass of cancer cells. A similar situation occurs with chronic myeloid >leukaemia, in which all the cancer cells have an abnormal chromosome, the Philadelphia chromosome, which again is a translocation that moves a proto-oncogene to a new position where it is de-regulated. Some experiments suggest that there must be two oncogenes involved, one to enable the cells to proliferate indefinitely (the 'immortalizing' gene) and the other to cause the cells to lose their own special shape and behaviour (the 'transforming' gene).

Genetic engineering

The ever-increasing amount that is known about gene regulation in prokaryotes (and in simple eukaryotes such as yeast) has become the basis for the fast-developing science of >genetic engineering. The organism most often used for genetic engineering is the laboratory workhorse *E. coli* and the enzymes that occur naturally in it. The latter are employed for 'snipping' DNA (when the enzymes are called >restriction endonucleases) and for sticking lengths of DNA together (when the enzymes are called ligases; *see* >ligation). Single genes or whole sequences can be put into the bacteria; this is called gene >splicing. The genes or gene fragments being inserted might be from another strain of *E. coli*, or they might be artificially produced DNA, or they might be genes from a different species (even a human being). The new DNA is called >recombinant DNA. But the recombined 'thing' in such a case – let us say that it is recombinant *E. coli*/human DNA – is only a piece of DNA; it is in no way a 'cross' between a bacterium and a human being.

Once a gene has been inserted into *E. coli*'s own DNA machinery (actually on a >plasmid rather than in the chromosome), it functions as an ordinary working gene, so that the bacterium finds itself making an alien protein. Thus human proteins can be produced by bacteria by means of a technique called >gene cloning, which is now in use on a commercial scale to make several useful proteins, such as insulin to treat diabetics.

The techniques of genetic engineering are also useful for improving strains of food plants, such as rice, wheat and maize. It is a feature of many plants that they can have doubled or tripled sets of chromosomes (polyploidy). If a useful gene, such as one that helps make a plant resistant to drought, is found in a wild species, it can be crossed into a good commercial strain by 'crossing in' a whole set of chromosomes. This technique has been used since the early days of genetics (Tshermak was a pioneer), but it is now much more sophisticated. As well as crossing-in useful genes found in the wild, genes from other species of plants can be spliced in by recombinant DNA techniques.

Genetic engineering as such has not yet been used to improve farm animals, though selective breeding has always been used and still is. Selective breeding is speeded up by techniques (biotechnological rather than genetic) such as >artificial insemination and >embryo transfer, the former used very widely particularly in the dairy industry, and the latter used sometimes when establishing herds of cattle or flocks of sheep of breeds new to a particular country. One technique that would be of overwhelming importance to cattle breeders has not yet been devised, even though the matter has been intensively pursued for over 50 years: this is the separation of X and Y sperm. If the sperm collected from a bull for artificial insemination could be separated, then the sex of the calf could be chosen, and all dairy calves would be female, and all beef calves male.

It is technically possible to >clone animals, and it has been done with sheep and mice, but the technique is not yet used for producing commercial farm stock, and is likely to remain too expensive in relation to its benefits, compared with the simple and effective technique of artificial insemination. In theory it might be possible to produce better farm animals by inserting single desirable genes into the genome, as is done with plants. However, gene transfer with animals has so far worked only occasionally, it being a hit-and-miss affair with more miss than hit. And so it will remain until the mechanisms of gene control in higher organisms are better understood.

Recombinant DNA techniques, though very useful, are also potentially extremely hazardous because of the risk that they could give rise to new forms of viruses or bacteria so virulent that there would be no cure for the diseases they caused and no possibility of wiping them out. (It is certain that such novel forms of bacteria and viruses are being developed for biological warfare, though no government will admit to it.) For this reason recombinant DNA research is subject to controls in all countries, though the controls and restraints are in most countries voluntary, not legally enforceable. Laboratory practice for such research is carefully controlled.

Genes cannot (at present) be created to order; the genes involved in genetic engineering as well as in conventional selective breeding must be ones that exist in real organisms. For this reason, it is extremely important to preserve all the different genes that exist in the wild and in cultivation. To take an extreme example, if all the apples in the world were Granny Smiths, and these were suddenly threatened by an infection that they were genetically susceptible to, there would be no chance of breeding resistant apples unless there was a source of non-Granny Smith genetic stock to select from. Modern specialized agriculture has already done too much in the way of narrowing the genetic range of commercial plants (and animals), but now 'gene banks' are being built up to prevent any more loss of genetic variety.

The Human Genome Project

By far the most ambitious research programme that the biological sciences have ever staged is the Human Genome Project (>genome means the total of the DNA, including both genes and >junk DNA). The plan is no less than to map the entire human DNA sequence base by base. There are four stages of investigation (though they will not necessarily be carried out consecutively).

1 By using linkage to already-known markers (many of them >restriction-fragment-length polymorphism loci), a linkage map with a resolution of about 1 centimorgan will be built up for all the chromosomes. At the outset of the project, such a map already existed but only with a resolution of 8–10 centimorgans.
2 A corresponding physical map of the chromosomes will be made, using observable morphological variants, so that the genes whose linkage relationships are known can be physically located, and their actual distance from one another assessed.
3 Each region of each chromosome will be studied by cutting it into short stretches with restriction enzymes and reproducing these as clones. Producing these clones is not difficult, though the number of separate clones is daunting; the problem will come in identifying the overlaps among many thousands of clones for each region and finding their correct order.
4 Once this has been done, the sequence of the overlapping clones will be analysed. This is the most labour-intensive and expensive stage of all because present techniques mean that it takes one person one year to sequence 100,000 base-pairs: at that rate the sequencing of the approximately 3 billion base-pairs in the human genome will take 30,000 person-years.

The challenge to present biological research capabilities is immense, and the challenge to information technology is almost as great, with existing database techniques having to be greatly expanded to cope with the analysis of vast amounts of overlapping data and with disseminating up-to-date information to research teams.

Appropriately, it is a global project. The idea started in 1985, when the molecular biologist Walter Sinsheimer wanted to found an Institute to Sequence the Human Genome at the Santa Cruz campus of the University of California. Many biologists were attracted by the idea of their science at last having the sort of money spent on it that is taken for granted in the physical sciences, and the project became the subject of top-level political manoeuvring in American research circles. It was fortunate that at the end of day leadership of the project was in the hands of James Watson, and equally unfortunate when internal wrangling led to his sudden resignation in 1992. International co-ordination is being managed by the Human Genome Organization (HUGO), headed by Sir Walter Bodmer. The work is parcelled out to research teams all round the world, each of which is assigned to work on one particular chromosome. The databases collating the results are in the United States, Britain, Germany and Japan. The total cost was at first estimated at $3 billion, but by 1990 this had increased to $10 billion.

The project is not without its critics. There have been three types of objection. The earliest was that it would absorb much of the money available for biological research, thus endangering work on more fundamental issues. After all, it was said, the Human Genome Project breaks no new ground technologically; it is just a huge data-crunching exercise. This criticism was made strongly by the American molecular biologist David Baltimore, a Nobel Prize-winner. Next, there were worries about the sensitivity of the project to commercial exploitation: could an international co-operative venture of pure research really work if at the end of the day there were potentially huge profits to be made somewhere? This came to a head in 1989 when the Japanese were slow to pay their contribution to HUGO: Watson wrote in the strongest terms (perhaps unwisely, referring to events in World War II) to say that the United States would not hesitate to withhold data from the Japanese unless they paid up. Finally, objections have been made on ethical grounds. Full knowledge of the human genome will mean that in principle all humans, born and unborn, can be screened for all genes, and the abuses that this could lead to are enormous. Research into ethical implications has been allocated 3 per cent of the project's budget.

At the same time that the Human Genome Project is absorbing so much money and effort in the geneticists' community, other similar projects of great importance are up and running. There is an independent British project, the object of which is to sequence all the genes in the human genome. This is a lesser task than tackling the entire genome, as it ignores all the sequences that are non-coding (>junk DNA), which actually comprise the greater part of the DNA in humans as in other higher eukaryotes. Much of the work is being carried out at the Medical Research Council's Molecular Genetics Unit, whose director Sydney Brenner was one of the prime movers of the research project. Co-ordination of work done throughout the country is carried out at the Human Genome Mapping Project Resource Centre at Northwick Park, near London.

Another project, devoted to the yeast genome, is being carried out in various laboratories in the countries of the European Community. The scale of the task is modest compared with the human equivalent, as the yeast genome is only 1/200 the size of the human one, but the genes are packed more closely, there being some 6,500 genes in yeast. The project is budgeted at ECU 20 million, and is due to be completed in 1995. A project to map the pig's genome has also begun, with 16 countries contributing and the work being co-ordinated at Edinburgh University. Although the pig's genome is comparable in size to the human one, some aspects of research can proceed faster owing to the facts that pigs have a shorter generation time, they have more offspring at one time, and there are no significant vetoes on experimental procedures.

3 GENETICISM: GENES AND POLITICS

From the time of the Darwin–Wallace discovery of evolution in the mid-nineteenth century, and even more since the discoveries of Mendel and Crick and Watson, there has been an almost supernatural belief in the power of the gene. Genes are invisible, but if science says that they exist and that they exert enormous influence, then they will be brought into practical human life. Belief in the power of genes has been termed 'geneticism', and it has been going strong for well over a hundred years, although it has changed the focus of its attention from time to time to keep pace with intellectual fashion. The coinage of the term is a recent one, and is somewhat unfortunate because of the possible implication that 'geneticism' is what geneticists do or think. As will become clear, this is far from the case. Geneticism is purely political, whether in the public or private sphere.

The curse on Ham

Geneticism is an idea as old as the Old Testament. In Genesis, Noah curses his son Ham, saying that Ham's descendants, i.e. the Africans, will be subservient to those of his other two sons. This curse was taken literally by Europeans as an excuse for the practice of slavery. Geneticism gave the idea a new impetus. Within a few years of the publication of *On the Origin of Species*, some very unpleasant scientific papers were being written, with the weight and prestige of Darwinism behind them (though not with Darwin's personal support). One of these was entitled 'The Negro's Place in Nature', and was written in 1866 by James Hunt, President of the Anthropological Society in London. Hunt said: 'The analogies are far more numerous between the ape and the negro than between the ape and the European.' In another paper he stated that 'there is as good a reason for classifying the negro as a distinct species from the European as there is for making the ass a distinct species from the zebra.' Hunt (later mourned in an American obituary as 'the best man in England') was wrong even by the scientific standards of his own time, it having been established that the ability to interbreed was the criterion for being in the same species. None the less, the idea of the evolutionary inferiority of the black races was extremely convenient to the European powers that had colonial empires, and it was almost universally believed. Hunt was only one of many writing in the same vein.

There are doubtless many white people left (and not all of them in the Republic of South Africa) who still have some idea that it is 'scientifically true' that black people are inferior. It is a great reproach to genetics as a whole that this should be possible, being more or less the equivalent of believing that the sun goes round the earth. Let it be remembered, however, that it is only twenty years since the most recent, deadly serious attempt by scientists to revive this myth. This was the infamous crusade in 1968, led by the American psychologist Arthur Jensen, to persuade the public that the relatively poor performance of black American children at school was due to their inferior genotypes (*see* IQ).

Eugenics

Another idea that came out of the intellectual upheaval after Darwin was that of eugenics, the 'science' of attempting to improve a human population by controlled breeding. The word itself was coined by Darwin's cousin Sir Francis >Galton, a gifted biologist and a keen student of human variation (one can hardly say genetics, as that science did not really begin until after Mendel was rediscovered). The newly prosperous British middle classes were happy to believe that they had succeeded in the economic struggle for existence (which, they thought, must be parallel to the one in nature depicted by Darwin) by reason of their superior genetic endowment. Most middle-class British men had large families from a sense of duty (and were very careful to keep from their wives the news that birth control was not only possible but easy), it being to the benefit of the country that their excellent genes should be reproduced as often as possible. (Observe that nationalism came into it, too: just as there was a struggle for existence between individuals, so also between nations; a superior Englishman's duty was to breed fine offspring in greater numbers than his German or French counterpart was doing.)

Before long a terrible fear struck the collective mind of this generally complacent class – fear of the genetic inferiority of the poor. What if the poor should outbreed them and swamp the race with their deplorable (as it must be, for science said so) genetic heritage? So the thrust of eugenics turned from the 'positive' eugenics of, say, Lady Galton having as many children as possible, to the 'negative' eugenics of stopping Mrs Smith from having any at all. (Galton himself only believed in positive eugenics, and cannot be held responsible for having started the negative eugenics ball rolling.)

'Bad' genes

Eugenics, and negative eugenics in particular, got an immense boost from the rediscovery of Mendel's laws. It was soon believed that there was a gene for low intelligence, a gene for improvidence, a gene for criminality (or was it several genes, one for robbery, another for rape . . .?). In perfect seriousness research studies on these theories were published, such as the one on the 'Jukes' and 'Kalliwak' families (both mercifully pseudonyms), which reported that a gene for shiftlessness ran through the pedigrees of both, with heterozygotes 'partially shiftless' and homozygotes 'shiftless'. This was perhaps geneticism's finest hour, but before laughing at it one must remember that the seriousness with which these results were presented and accepted was exactly the same as that obtaining in scientific research now. The shiftlessness study was carried out in the United States, and it was in the 'Land of the Free' that Mendelian genetics was most keenly taken up as the basis of programmes of negative eugenics. By 1910 no fewer than 15 states had passed laws allowing 'unfit persons' to be sterilized.

The only survival of the idea of positive eugenics was in the suggestion put forward in the 1930s by the American geneticist H. J. >Muller that sperm banks should be established so that women could choose to have the highest possible quality of baby. Unfortunately there could not be a guarantee that the Great and the Good whose sperm would be banked would not have just as many deleterious recessive genes concealed in their genotypes as any of the rest of us, as Muller realized, so the prospect of improving the human gene pool by this method was in doubt from the start. But what it was all about, of course, was not improving the gene pool but having brighter children. Intelligence has always been a subject that attracts rampant geneticism, but even so, this idea – that the genotype is so overwhelmingly influential that even the artificial insemination of a middling woman offering a baby a middling environment will still produce the desired result – is extreme.

In the 'post-DNA' era (that is, since Crick and Watson's launching of molecular genetics) eugenics has focused on the elimination of particular deleterious mutations. The method used is usually selective abortion of foetuses that can be identified as having (or as likely to have) one of these mutations, and the list of genetic conditions that can be detected by >prenatal diagnosis runs into dozens (all major chromosomal abnormalities are included, and the number of single-gene effects that can be diagnosed increases all the time as techniques become ever more sophisticated.) In some cases, and particularly in late-onset diseases (e.g. Huntington's chorea) where it would be impossible to diagnose the foetus as affected, linkage to a known genetic marker may indicate such a high probability of the mutant gene being present that abortion is considered to be justified.

In the case of some recessive mutations it is easy to test adults to see whether they are carriers. One of these is thalassaemia, a type of anaemia resulting in a deficient protein in the red blood cells, which still occurs with high frequency in some Mediterranean countries, in communities elsewhere of Mediterranean descent, and among black Americans. The frequency is high enough in these places for it to be worthwhile screening all pregnant women and, if they are found to be carriers, then testing their partners as well; if the man is also a carrier, the foetus has a one-in-four chance of being a victim of thalassaemia. Programmes of prenatal testing and selective abortion along these lines have been extremely successful in reducing the frequency of thalassaemia in Sardinia, Italy and Cyprus and in the Cypriot community in London, with the incidence of thalassaemic babies falling from around 5 or 6 per 1000 to less than 1 per 1000 in the course of three or four years. In the London project, an estimate has been made of the relative cost of the screening programme compared to the cost of treatment of thalassaemic patients; it seems that the prenatal diagnosis service (excluding the cost of abortions, which was not calculated) can be run for one year for less than the projected cost of treatment for a single patient over 20 years.

There can be no doubt of the cost-effectiveness of this approach, and that is one reason why it is used in practically every country that has the medical resources (which are quite modest). Some countries have religious principles that do not allow abortion for whatever reason, and they do not carry out programmes of this sort. Catholic countries seem able to accept the abortion of foetuses with genetic disease, as in some of the thalassaemia examples, but Islamic ones do not. But the moral acceptability of aborting foetuses on account of their genotype is far from universally accepted, even in countries where it is often done. Many people are worried about the arbitrariness of deciding which genes shall cause carriers to be considered unfit to live. An African doctor voiced this eloquently in a letter to the *British Medical Journal*, reporting that his condition of having six digits on each hand and foot was luckily thought of as neutral by the elders of his tribe, though in a neighbouring tribe it was believed to be unlucky and any baby born like this was killed. It is easy to say that it is obvious to a modern society that something like extra digits does not matter, but the questions remain: Who is to decide? And on what criteria? How should the decision go in cases of a late-onset disease such as Huntington's chorea? The individual will have about 40 years of normal life, a good run by any reckoning. He or she need not pass on the gene to offspring, since alternative reproductive strategies are available. And during his or her lifetime a cure for the disease may be found.

Reproductive technologies

The other side of the coin of deciding what babies *not* to have is deciding what babies one *will* have. A great deal of attention has been focused over the last decade on the plight of the infertile. The problem is not new, nor is it uncommon: it is estimated that in European populations about 10 per cent of marriages are infertile, due to deficiencies in the reproductive system(s) of one or both partners or to a reproductive incompatibility between them. The solution used to be either adoption or resignation. Adoption is still available, but there have been far fewer babies for adoption since legalized abortion became more available. Resignation has given way to the idea that there is a 'right' to have a baby, now that medical technology is offering more and more ways round the obstacle.

And here geneticism re-enters the picture: what one is considered to have is not just the right to a baby but to one's 'own' baby – a baby with one's own genes in it. Some technologies, notably *in vitro* fertilization of the woman's egg with the man's sperm, allow a baby to be born that is as much genetically related to its parents as a naturally conceived one is. Others, for example egg or sperm donation, mean that the baby has half its genes from one of its parents and the other half from someone else. Even this is thought to be a great improvement on an unrelated baby.

The biological fact is that of the 100,000 or so genes in the human genome, about 90,000 are non-varying. In other words, the vast majority of one's inheritance is of genes that say, 'Make a human being,' not 'Make this particular human being.' Even if your son is genetically the sperm donor's and not your own, 90 per cent of his paternal genes will be absolutely identical to yours. Yet people feel very strongly about the other 10 per cent. The biological father in the 'Baby M' surrogacy case in the United States in 1987 pleaded his need for his own child because his genes were the only ones of his family to have survived the Holocaust and could not be allowed to die out now. (The genes had, of course, survived anyway, in that he had fathered a baby.) The issue was whether he should be allowed to have the child when the surrogate mother wanted to keep her; he won. This story brings out another aspect of geneticism, namely the idea that it is important for a child to be brought up by a biological parent, or at least to know who their biological parent is. This form of geneticism gives rise to the belief that it is right for an adopted child to learn the identity of its 'real' mother, or even for a child conceived by artificial insemination to know about its 'real' father.

This insistence on the importance of one's own genes is in direct contradiction to contemporary attitudes towards abortion. If a person's own genes matter so much, how come it is all right to abort the package they are in at this moment? This contradiction was tested in the English courts in 1987; a student whose girl friend was pregnant tried to prevent her from having an abortion, saying that the child was genetically his and that he had a right to have it if he wanted. He lost (having gone as far as the House of Lords), and public opinion was strongly against him. He had geneticism on his side, and exactly the same geneticism (the belief in the importance of paternity) as invoked by the father of 'Baby M', but in a less fashionable case.

Inequality

The biologist and socialist J. B. S. >Haldane once wrote: 'Any satisfactory political and economic system must be based on the recognition of human (genetic) inequality.' It is true. Whatever your political perspective, you must take into account, in your mind and in your dealings with your fellow humans, that we are not all quite the same genetically. By and large this matters staggeringly little, and to believe otherwise is to subscribe to one or other of the forms of geneticism described above.

But in a few instances it matters a great deal. There are some genetic conditions that alter drastically the quality of life for people who have them: a humane society must take account of this. This creates a tension between civil liberties and the protection of people who, for genetic reasons, are not able to take care of themselves or make decisions for themselves. There is also an obligation on everyone to think for themselves how far they are prepared to let their society go with eugenic measures, or in reproductive technology and research, or the genetic manipulation of other species. None of these is a question for 'experts', for there are no experts on what is right and what is wrong. But it is important that everyone should understand how genetics works, in order to be able to understand the choices that technology and medical science are now laying before us.

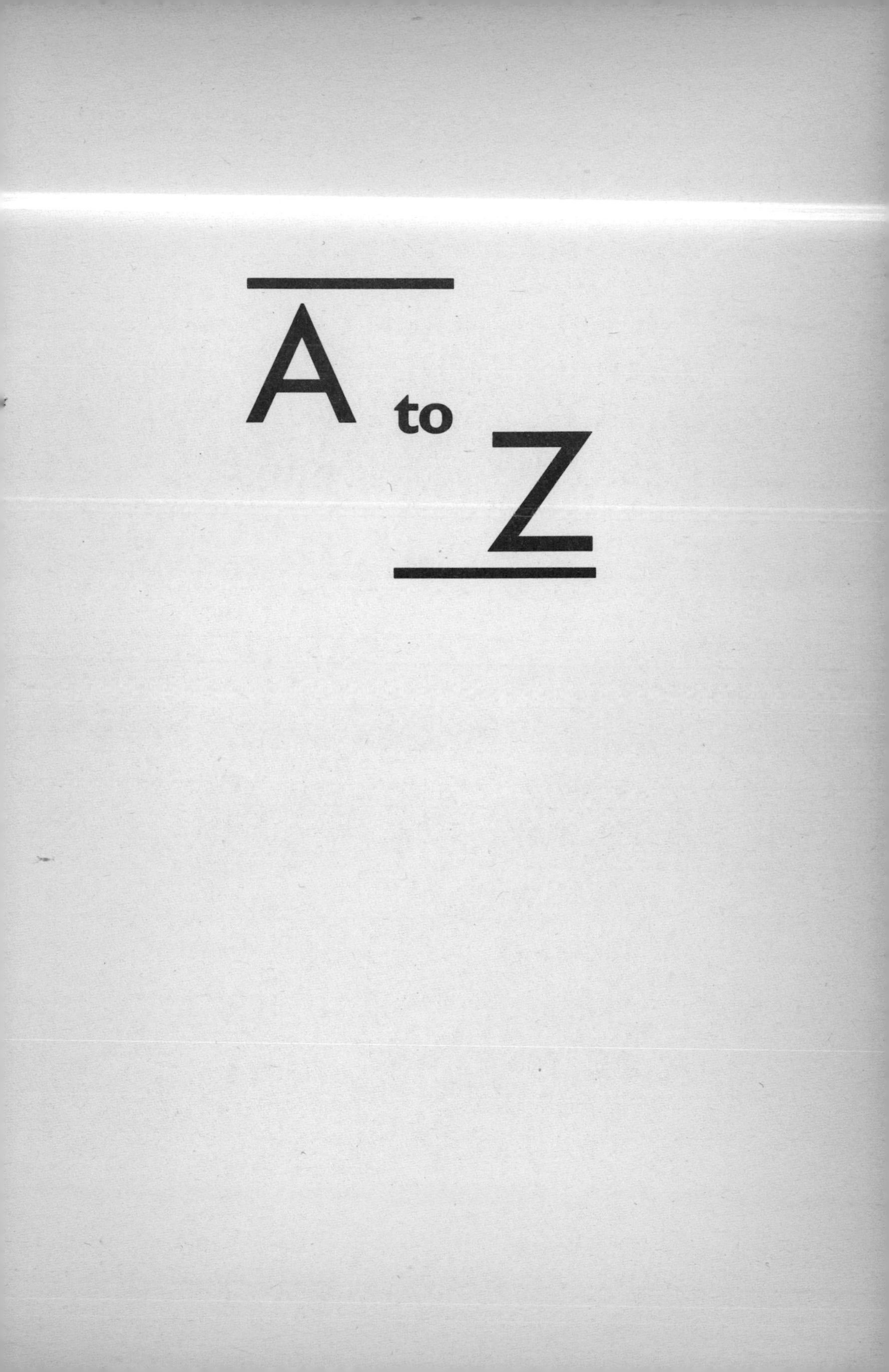
A to Z

A

A chromosomes, the normal chromosomes in a eukaryote. Occasionally extra chromosomes occur, which are known as >B chromosomes.

ABO blood group system, one of several genetic systems of antigens found on the surface of red blood cells in humans. It is medically important, because in a blood transfusion the donor's ABO blood group must be compatible with the recipient's. There are three alleles involved: G^A, G^B, and G^O; G^O is recessive to both the others, while G^A and G^B are codominant. The blood groups of the six possible genotypes are therefore as follows:

G^AG^A = group A
G^AG^B = group AB
G^AG^O = group A
G^BG^B = group B
G^BG^O = group B
G^OG^O = group O

The alleles G^A and G^B code for making an antigen, which is what is known as A or B; G^O makes no antigen. People with blood group A have in their bloodstream antibodies to B (known as anti-B), and people with blood group B have antibodies to A (anti-A); people with blood group O have antibodies to both A and B; and people with blood group AB have neither of the antibodies. In a blood transfusion, the person's circulating antibody would attack any incoming blood that had the corresponding antigen, setting off an adverse reaction within the bloodstream that could be fatal. Thus it would be dangerous to give group A blood to a person with blood group B (and therefore with anti-A antibodies). It follows that group O blood can be given in transfusion to a person of any type, as there are no

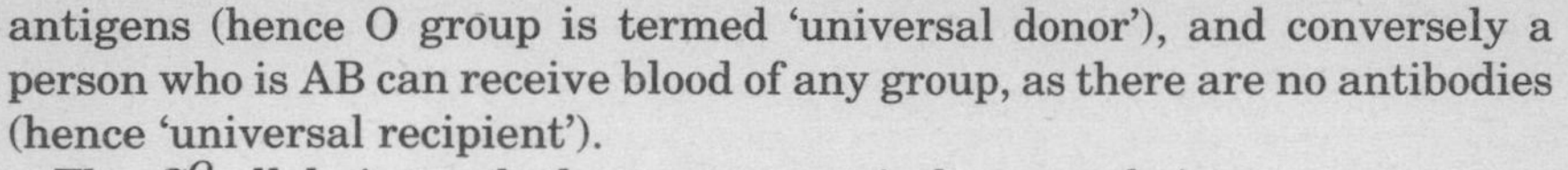

antigens (hence O group is termed 'universal donor'), and conversely a person who is AB can receive blood of any group, as there are no antibodies (hence 'universal recipient').

The G^O allele is much the commonest in humans, being at an average global frequency of 62 per cent. European populations are about 68 per cent G^O, 26 per cent G^A and 6 per cent G^B. The place with the highest frequency of G^B is north India, with about 30 per cent; this allele is virtually absent from Amerindian populations.

Abortion, the expulsion of the foetus from the uterus. *Spontaneous abortion* (commonly known as 'miscarriage') is quite common, occurring in perhaps as many as 20 per cent of human pregnancies, particularly within the first two months. About half these foetuses have chromosome abnormalities. Some diseases cause abortion, e.g. brucellosis in cattle, and many species will spontaneously abort if disturbed; for example, the smell of a male mouse other than her mate causes a female mouse to lose her litter. *Induced abortion* of normal foetuses is very widely used as a method of birth control, legally in about half the countries of the world and illegally elsewhere. *Therapeutic abortion* is practised in advanced countries to prevent the birth of deformed or diseased foetuses. (The word 'therapeutic' is curious in the context, as it means curative: the mother cannot be the person cured, as she was not ill, and neither can it be the foetus, who ends up dead.) In 1988, nearly one in five pregnancies in Britain were terminated, and of these 180,000 abortions, some 5 per cent were performed because of the risk that the foetus might be abnormal. *See also* selective reduction; and pp. 44–5.

Acceleration, the appearance of a structure in >embryogenesis earlier than in ancestral forms. The opposite is >retardation.

Acceleration, rule of, the rule that in an organism's development the structures that are formed first are the ones most important to the organism's overall function.

Acentric chromosome, a chromosome with no >centromere. This may be merely a large part of a chromosome that has broken off from another one. Acentrics tend to get lost because they cannot be manoeuvred during the normal cell-division process.

Acetabularia, a single-celled green alga; a eukaryote (unlike the blue-green algae), with a diploid chromosome number of about 20. Because it is so large (about 2/5 in/1 cm long) it is useful for experiments in which a cell's nucleus is removed and implanted into a denucleated specimen of a different strain or species, thus clarifying the role of chromosomal versus >cytoplasmic inheritance.

Achiasmate meiosis, a >meiosis in which there are no chiasmata (*see* chiasma), i.e. no >crossing-over takes place. This happens in some species, including *Drosophila melanogaster*, where only the homogametic sex has crossing-over.

Achondroplasia, a human disease caused by a dominant allele. Easily recognizable at birth, the features of the disease are very short limbs and a large head (but usually a small nose); adults grow to a height of 4 feet (1.2 m) or less. Achondroplastics are perfectly healthy and have normal intelligence, fertility and so on. About one baby in 50,000 is born with achondroplasia. Some 80 per cent of them have normal parents, which means that the achondroplasia gene must have arisen by a new >mutation in these cases, which seems to happen slightly more often when the father is over the age of 30. People with achondroplasia tend to marry others with the same condition, i.e. the marriage is between heterozygotes (*Aa*, where *A* is the dominant allele for achondroplasia and *a* the corresponding normal one). Their offspring are therefore expected to be $\frac{1}{4}$ *aa* (normal), $\frac{1}{2}$ *Aa* (achondroplastic) and $\frac{1}{4}$ *AA*. These last are not in fact the same phenotypically as their *Aa* parents. The *A* allele is strictly speaking semi-lethal, with the homozygote more severely affected than the heterozygote; *AA* babies die shortly after birth.

Acquired characteristics, inheritance of, a mistaken idea that has dogged genetics for years and still persists; see pp. 9–10. It seems to stand to reason that if an animal acquires at some point during its lifetime a trait that is useful or that helps it to survive, it will pass it on to its offspring (and that in this way evolutionary progress will be made). It is plausible, but it does not happen. Wallace, co-proposer of the theory of evolution by natural selection, wrote an excellent refutation:

> 'Neither did the giraffe acquire its long neck by desiring to reach the foliage of more lofty shrubs, and constantly stretching its neck for the purpose, but because any varieties which occurred among its ante-types with a longer neck than usual *at once secured a fresh range of pasture over the same ground as their shorter-necked companions, and on the first scarcity of food were thereby enabled to outlive them.*' (Wallace's italics.)

Let us look at Wallace's example in genetic terms. Even if a giraffe could make its neck longer by stretching, its offspring would have a normal-length neck, and each would have to do its own stretching to get a longer one. If there was a gene *n* for neck length, all giraffes were of genotype *nn*, and all their offspring would be *nn*, normal-length neck. But suppose a mutation arises, *N*, which is dominant over *n*, and gives a longer neck; then the giraffes that carry this gene (genotype *Nn* or *NN*) are longer-necked than the normal type, with the result that Wallace describes.

The point is that there is no way in which anything that happens to the individual giraffe's neck can be incorporated into what it passes on to its descendants. The only genetic route from one generation to the next is via the DNA in the chromosomes, and this DNA is always (barring the very

rare occurrence of mutation) an exactly faithful copy of the DNA in the preceding generation.

The key figure in disproving the inheritance of acquired characteristics was the German zoologist August >Weismann. He wrote several impeccably argued treatises (based on theory rather than on experimental results), including *Das Keimplasma* (1892, trans. *The Germ-Plasm: A Theory of Heredity*). His insistence that heredity must reside in units unaffected by what happens to the individual's own body was soon supported by Mendel's findings, which came to light in 1900. Since that time the idea of the inheritance of acquired characteristics has been rejected by scientists (but see Trofim Denisovitch >Lysenko for an unfortunate exception).

Acrocentric chromosome, a chromosome whose >centromere is towards one end. If the centromere is right at the end, the chromosome is called >telocentric.

Acromegaly. *See* growth hormone.

Active immunity, immunity to a disease that has been acquired through previous exposure to it. *See* immune system; passive immunity.

Adaptation, the acquisition of characteristics that increase an organism's reproductive success in a particular environment. Adaptation can happen to an individual organism if it carries a favourable mutation, but in evolution it is a property of populations, not individuals. From the evolutionary point of view it is only reproductive success that counts, not the efficiency or otherwise of the animal's interaction with its environment. *See also* acquired characteristics, inheritance of.

Adaptive radiation, the evolutionary diversification of a number of species from a common ancestral species, made possible by the appearance of vacant ecological niches into which new species can move. This can happen after a catastrophe or a colonization.

Adaptive value, the amount of increase or decrease in reproductive >fitness (relative to other genotypes and in a particular environment) that a given genotype or mutation confers.

Additive variance, the component of >genetic variance that is due to the additive effect of two or more genes on the trait in question, i.e. excluding any dominance or interaction.

Adenine (abbreviation: A), one of the four nitrogenous bases which form the core of both DNA and RNA. It pairs with thymine (T) in DNA and uracil (U) in RNA. *See also* replication.

Adenosine deaminase (ADA), an enzyme involved in the immune system. Lack of it is one of the causes of >severe combined immune deficiency.

Adenoviruses, a group of viruses which cause certain respiratory diseases (e.g. laryngitis and some kinds of pneumonia) in humans and other mammals. Their genome consists of double-stranded DNA but contains only about 12 genes.

Adoption, the raising of a baby by a person or persons not its biological parents. It is frequently found in non-human species; orphaned infants are not usually left to die among animals that live in social groups. A macaque monkey mother who has lost her own infant will even kidnap another as a replacement. Legal adoption was formerly a common means of providing a childless couple with a child, or an important man with an heir (such as Augustus Caesar, adopted by Julius), but it now occurs far less frequently in developed countries. The availability of abortion has drastically reduced the number of candidates for adoption (causing some couples to turn to the Third World as a source of babies), and the new reproductive technologies have given other possibilities to childless people. Adopted children have sometimes been used in the investigation of questions to do with >nature versus nurture, such as the heritability of intelligence (*see* IQ) or obesity; theory predicts that, if environmental influence is more important, they will resemble their adoptive parents more than their biological ones, and vice versa if genotype is more important.

Agamospermy (from the Greek *a*, 'without', *gamos*, 'marriage', *sperma*, 'seed'.) a process of seed formation in plants in which fertilization does not take place. *See also* apomixis.

Ageing, the phenomenon of progressive failure of self-repair of an organism. The process is not understood. The various theories include: build-up of >somatic mutations, age as a non-specific >auto-immune disease, and increasing inefficiency of DNA repair leading to >error catastrophe. These are all physiological explanations, but the >disposable soma theory is based on the population geneticists' view of a species' total biological investment. *See also* tissue culture.

Aggregation chimaera, an organism artificially created *in vitro* by mixing cells from two genetically different early embryos. The embryo is then reimplanted into a surrogate mother. The individual is a >chimaera, as can easily be seen if two very different strains are used, e.g. one black and one white mouse. *See* tetraparental mice.

Aggression, any behaviour intended to frighten or harm another. There has been heated discussion (decorated with such clichés as 'nature red in tooth and claw') as to whether aggression is genetically determined. Darwin felt that aggression within and between species must be part of the struggle for existence and hence a necessary feature of evolution.

In his book *On Aggression* (English edition 1966), Konrad >Lorenz proposed that it is natural, occurs in all species and is genetically determined. He also suggested that in most species there are (genetic) mechan-

isms that make the individual direct its aggressive behaviour outside its own group, and that in species which have the most dangerous natural weapons (fangs, claws, etc.), there are highly developed inhibitory mechanisms to cause individuals to use bluffing, submissive gestures, etc., rather than actually fight and harm one another. He went on to say that the problem with humans was that they were not naturally armed and therefore did not have these inhibitory mechanisms; so once they had armed themselves with artificial weapons, there was nothing to stop them killing one another.

Lorenz's facts were challenged by the sociobiologist E. O. Wilson, who noted that in species that have been observed for long periods in the wild aggression leading to serious injury or death is often seen: hyenas, lions, macaque monkeys and chimpanzees all kill members of their own species, and sometimes eat them too. Noting that these incidents only become apparent when a species has been observed for at least 1000 hours, Wilson remarked that one murder per 1000 hours is a great deal of violence by human standards. Human aggression is therefore neither unique nor extreme. Sociobiologists believe that aggressive behaviour, including both fighting to the death and ritualized conflicts, is genetically determined, and using >game theory analysis, they calculate the evolutionary advantage to the individual of handling aggression by fighting or by submissive behaviour. *See also* sociobiology.

Agouti, a coat colour in mammals. The coat appears dull brown, but actually each hair has bands of black and light brown. It is the wild-type colour of mice, rabbits, etc., and is controlled by a dominant allele, *A*, in a series with numerous other alleles. Variation at the agouti locus is responsible for many of the fancy colours that hobby breeders concentrate on. The word itself is from Guarani, a Brazilian language.

AIDS, acquired immune deficiency syndrome, a viral disease affecting humans and closely related species, in which the body's immune system becomes inoperative so that the patient ultimately dies of infection by some other disease. The agent that causes it was identified in 1983 as the human immunodeficiency virus (HIV), which can only be passed from one person to another when blood, semen or vaginal fluids pass directly into the bloodstream. This can happen during sexual intercourse or as a result of sharing needles for drug injection; and if a donor of blood, semen, breast milk or an organ for transplantation carries the virus, the recipient may become infected. (However, donated blood and semen are now both rigorously treated to avoid this possibility.) Not everyone who has the virus in their bloodstream will develop AIDS, and it is not true that homosexual men are particularly susceptible to either the virus or the disease. The virus is passed on in exactly the same way by AIDS patients and by HIV-positive people (those who have the virus but no symptoms).

AIDS has no genetic component, but there is confusion about this, as a

mother who has the virus can transmit it to her baby during pregnancy or birth, or in her breast milk. *See also* severe combined immune deficiency.

Albinism, a genetically determined condition in which there is no pigmentation in the skin, hair, eyes, etc. The cause is a recessive allele; the corresponding normal allele codes for the enzyme tyrosinase which is one step in the production of >melanin, the pigment that usually gives these tissues their colour. The albino gene in humans is the same as the one in mice, rabbits and other animals, and was one of the very first traits to be identified as showing Mendelian inheritance in humans (reported in 1903). Albino humans are healthy and not at a disadvantage except that they must avoid bright sunlight which damages their skin and eyes. There is an interesting group of North American Indians, the Hopi, who have a relatively high incidence of albinism; this is believed to be due to the fact that albino men and women spend more time inside (and consequently reproductively busy) than their normal comrades. Very occasionally, a different mutant causing albinism arises which is dominant.

Algeny, >genetic engineering. A word formed to be like 'alchemy', algeny includes any sort of manipulation of genetic material *in vivo* or *in vitro*.

Alkaptonuria, a human disease caused by a recessive allele. The actual mutation is a failure to produce an enzyme that converts homogentisic acid into urea to be excreted in the urine; thus the urine contains homogentisic acid, which turns black. There is sometimes also degeneration of cartilage (especially in the spinal discs). This was one of the >inborn errors of metabolism identified by Sir Archibald >Garrod.

Allele (from the Greek for 'each other'), one of the variant forms of a gene. An individual has two alleles for each gene, one on each of the two homologous chromosomes, at the appropriate locus for that gene. If the two alleles are identical, the individual is a >homozygote (in respect of that locus), and if they are different, the individual is a >heterozygote. Each gene may have a very large number of alternative alleles, since an allele need only differ from the original form by one base-pair out of thousands (*see* mutation). The allele that is the original form is sometimes known as the >wild type. *See also* dominant; recessive.

Allele-specific oligonucleotide, an artificially produced sequence of nucleotides that is known to be exactly complementary to a particular sequence of >DNA. These sequences, usually about 18 >base-pairs long but sometimes as long as 80 base-pairs, can be used to detect the presence or absence of the sequence in question (perhaps a disease-producing mutant) in a sample of natural DNA: when a quantity of the nucleotide is added to the natural sample, it will bind if it can find its exact complementary sequence, but not otherwise, being accurate to a single base. In a case such as sickle-cell anaemia, where the exact sequence of the mutation site

is known, an allele-specific oligonucleotide is an efficient way of detecting the presence of the mutation, and is used in prenatal diagnosis.

Allelic exclusion, the fact that the lymphocytes of an individual who is heterozygous at one of the >immunoglobulin loci only make one of the immunoglobulin types. The mechanism by which the other allele is prevented from being active is not known; it is not a simple case of dominance.

Allelic series, the set of alleles available for a given locus. They can be listed in order of >dominance, though this order might not be a simple hierarchy and there might be some >co-dominance. An example known to mouse breeders is the >agouti series, with five alleles in use in the fancy out of some 20 known ones.

Allergy, the body's response to a substance to which it has been previously hypersensitized, by contact with a particular antigen (known as the allergen). If the allergen enters the gastrointestinal tract, the symptoms are vomiting or diarrhoea; if it enters the respiratory tract, the result is sneezing, runny eyes and inflammation of the membranes; and if it makes contact with the skin, there are rashes or swellings. Almost any molecule can act as an allergen, although some – such as pollen and animal fur – are particularly common. There may be a genetic basis to an individual's liability to become allergic to particular substances. Some patients respond to immunotherapy (repeated small doses of the allergen to alter the hypersensivity response); others are treated with antihistamine.

Allometry, the differential growth of one part of the body compared to another or to the body as a whole. It is one of the most frequent phenomena in evolution (e.g. the positive allometric growth of the human cranium compared with that of the immediately ancestral form), and is perhaps under very simple genetic control: all that is needed is to keep a given gene switched on for longer than was normal.

Allomones, a group of chemicals that act as chemical messengers between members of different species, in the same way as >pheromones do within species. An interesting example is the substance secreted by aggressive species of ants to frighten and subdue other species of ants that they intend to capture as slaves.

Allopatry, the occupation by populations of geographically separate ranges. The formation of new species can occur if two groups of the same species become geographically separated and develop divergent changes in their gene pools: this is allopatric speciation. *See also* adaptive radiation.

Allophene, an organism that has received a transplant of foreign cells. The transplanted cells have their own genotype different from the host's, and they make their own gene products. An example is a patient who receives a bone-marrow transplant to replace their own defective marrow tissue. The word is also used for >tetraparental mice.

Allopolyploid, a >polyploid organism in which the chromosome sets have come from more than one species. *See Triticum aestivum*; autopolyploid.

Allozyme, one of the variant forms of an >enzyme coded for by alleles at a given locus. *See* isoenzyme.

Alternation of generations, the system by which plants have a life cycle consisting of a haploid phase and a diploid phase. The diploid stage, which is the 'plant' itself as visible, makes male microsporocytes in the anthers; these divide by meiosis and each produces four microspores. These are the male gametophytes, and are haploid. At the same time the plant also produces female megasporocytes (diploid) which undergo meiosis to give rise to megaspores (haploid): these are the female gametophytes, and they go through some mitotic divisions (*see* mitosis) but still remain haploid. Fertilization unites a megaspore and a microspore to form a diploid zygote, and the sporophyte phase begins again at that point. In higher plants the sporophyte is much the longest phase, the gametophyte being a transient phase involving a few microscopically small cells.

Altruism, behaviour by one organism that benefits not itself but another. More specifically, from the genetical point of view, it is behaviour that increases the reproductive performance of another. It has been a puzzle to evolutionists to explain how altruism could have evolved, since it appears to be contrary to the way natural selection works (i.e. the genotypes that are 'selected' are those that leave most offspring), but it is undeniably common in nature.

There are three categories of altruism:

- *straight altruism* – e.g. lower-ranking wild turkeys who assist their brothers in inter-group fights over females with which they themselves will not be allowed to mate.
- *compulsory altruism* – e.g. the killing of one offspring to feed it to the others.
- *reciprocal altruism* – 'one good turn deserves another', a much rarer phenomenon but sometimes found in bands of baboons which beg food on one occasion and repay it on another.

It must be emphasized that there is no suggestion that altruistic behaviour is in any way conscious; indeed, it is a tenet of sociobiology that altruistic behaviour (like all other sorts of behaviour) is directly dictated by the genes.

Altruism was supposed by some to have evolved because such behaviour was of the benefit to the group (*see* group selection), but it has been shown that group selection does not work mathematically. The current thinking in sociobiology is that altruism evolved through >kin selection because by sacrificing its own reproductive performance the altruist increases the number of its genes (i.e. identical copies) transmitted by its relatives. Even this has been strongly criticised by geneticists because of the closeness of the relationships needed to make the mathematics work out. One particular case in which kin selection could explain altruism is in the >social insects. In humans it is possible to imagine how kin selection could explain such altruism as a father going without food to feed his children, but it is nowhere near explaining the mass altruism seen in world wars and other disasters. This anomaly is admitted by those sociobiologists who agree that culture makes humans a special case (which is by no means all of them).

Alu sequence, a sequence of highly repetitive DNA in the human genome. It is about 300 base-pairs long, and is repeated 300,000 times, the copies being scattered around the chromosomes. The Alu sequence copies make up some 3 per cent of the total human DNA. The sequence does not seem to have any genetic meaning, but as it is somewhat related to the sequence of one type of RNA, it is possible that the Alu sequence arose by the action of >reverse transcriptase and hence could have originated from >retroviruses. The copies exist in slightly varying forms, and it is the combination of these variants that gives the individuality analysed in >genetic fingerprinting. The Alu sequence got its name from that of the restriction endonuclease, *Alu I*, by which it was isolated.

Alzheimer's disease, a human disease existing in several forms, one of which may be genetically determined. The symptoms are mental confusion and loss of memory, often with loss of speech and lack of muscular control. Brain scans show deposits of plaques on the nerve cells, which are made of the protein beta-amyloid, and it is an abnormality in this protein that is suspected of being the cause of Alzheimer's disease. Most people who get Alzheimer's disease do not develop it until they are over the age of 70. The early-onset form (developing at about 50) is the one that has a genetic component, the gene responsible having been located on chromosome 21. The late-onset form, which develops when the patient is over 70, is linked to two markers on chromosome 19. The exact mode of inheritance has not been worked out, owing to uncertainties in the interpretation of family histories when most of the people who may have been affected by the disease are already dead and may not have been correctly diagnosed one way or the other.

Amber codon, one of the three >nonsense codons in DNA that code for 'stop' and therefore signal the termination of the protein being synthesized. Amber is UAG. *See* protein synthesis.

Ames test, a technique for testing whether a substance is a >mutagen. A culture is made consisting of a strain of *Salmonella* that cannot synthesize the amino acid histidine but must get it from its food source. This strain is set in a nutrient medium lacking histidine, and the substance to be tested is added; if the latter is mutagenic, there will be some mutations of the *Salmonella* enabling it to make its own histidine and therefore to survive; if the substance is not mutagenic, nothing will happen.

Amino acids, the sub-units out of which proteins are made. There are 20 that commonly occur in animals, and about 100 more that are rare and found only in plants. Most amino acids can be synthesized in the body, but some, known as the 'essential amino acids', must be got from the diet. The essential amino acids for humans are: isoleucine, leucine, lysine, methionine, phenylalanine, threonine, tryptophan and valine. The list is a little different for other species. *See* genetic code; protein synthesis.

Amish, the, or the Old Order Amish, a religious community living in three areas of the United States. Descended from a small group of eighteenth-century German emigrants, the Amish have kept themselves apart from the rest of the American population in order to preserve their puritanical beliefs. They refuse to use any machinery, and they work their farms with hand or horse-drawn implements.

The Amish are an ideal subject for genetical research: the population size is small (about 50,000) and marriage with outsiders rare, so that there is much inbreeding, which can be traced through the excellent family records that the Amish have always kept. Several genes have been found at high frequency (due presumably to genetic drift), one being a recessive causing the formation of six digits on the hands. It was also among the Amish that the first evidence was found for a gene causing >manic-depression.

Amitosis, a form of cell division in which two new cells are formed without >mitosis. The nuclear membrane simply pinches itself in two, with equal numbers of chromosomes in each half; the genetic composition of the two new cells is therefore not the same. This happens in species that are highly >polyploid.

Amniocentesis, the drawing off of some of the amniotic fluid (the fluid in the sac surrounding the foetus). This fluid contains some cells from the foetus, and these can be cultured for >prenatal diagnosis of genetic defects. The procedure involves putting a needle through the skin of the mother's abdomen and through the wall of the uterus. It cannot be carried out until at least the twelfth week of pregnancy, and then several weeks are needed for tissue culture and tests, so that the results are not known until about the sixteenth week at the earliest. There is a 0.5–1.0 per cent risk of losing the foetus from the procedure itself. *See also* chorionic villus sampling; polymerase chain reaction.

Amphimixis, sexual reproduction in plants, including both self-fertilization and cross-fertilization.

Anaemias, a group of diseases of the blood, several of which are inherited. *See* glucose-6-phosphate dehydrogenase deficiency; sickle-cell anaemia; thalassaemia.

Anaphase, one of the phases of the two types of cell division, >meiosis and >mitosis.

Aneuploidy, the condition in which the number of chromosomes in a cell is not an exact multiple of the >haploid number. This can be either because one or more chromosomes have been lost or because of the presence of extra chromosomes. One form of aneuploidy is >trisomy, in which there are three copies of a chromosome: a well-known example is Down's syndrome, also called trisomy 21, in which there are three copies of chromosome 21.

Angiosperm, (from the Greek *angeion*, 'vessel', and *sperma*, 'seed') a plant that has an ovary to contain its seeds. Angiosperms comprise the >flowering plants, and are contrasted with the >gymnosperms.

Animal, any living thing that is capable of voluntary movement; but in popular use, generally refers to a mammal. Animals (especially in the latter sense) are used in scientific experiments in enormous numbers, and genetic experimenters have been among the heaviest users, where large numbers of results are needed for statistical analysis. In her book *Animals and Why They Matter* (1983), the British philosopher Mary Midgley quotes the following (referring to an American TV programme):

> [The interviewer] asked the scientists whether the fact that an experiment will kill hundreds of animals is ever regarded by scientists as a reason for not performing it. One of the scientists answered, 'Not that I know of.' [The interviewer] pressed his question: 'Don't the animals count at all?' [A second scientist] replied, 'Why should they?' while [a third] added that he did not think that experimenting on animals raised a moral issue at all.

Most of the experimentation in genetics is done with flies, plants, funguses or bacteria, and the moral acceptability of using such living things in this way is taken for granted. *See also* anthropomorphism; LD50.

Animalia, one of the five kingdoms of living things. *See* classification.

Anisogamy, the production of two unequal-sized types of >gametes in a species. It is the system used by all higher eukaryotes, in which the ovum is much larger than the sperm. The opposite is >isogamy.

Ankylosing spondylitis, a human disease probably caused by a virus but with an element of genetic susceptibility. The symptoms are the fusing together of the vertebrae of the spine, starting from the lowest, but even though there may be almost no movement in the back, the patient is usually well enough to lead a normal life. Men are affected four times more often than women, and there is often more than one case in a family. The factor that makes someone liable to develop ankylosing spondylitis seems to be linked with a gene in the >HLA complex, part of the immune system: almost all cases of the condition are in people with HLA type B27, though this type occurs in only 3 or 4 per cent in the population. The HLA genes may be directly involved in the disease, in causing an attack on the body's own tissues, or they may only be genetic markers linked very closely to the real genetic factor.

Annealing, the reassociation of complementary strands of DNA after they have been denatured (usually by melting at a high temperature); *see* denaturation. The annealed strands can either be original complementary strands or they can be any two strands that have enough base-pairs matching for reassociation to occur. This hybridization of DNA is one technique used in >genetic engineering. Annealing can also occur between a strand of DNA and a strand of RNA.

Anthropomorphism, the attribution of human characteristics to animals, things or deities. It is an idea that, if put into reverse, is useful in thinking about the evolution of humankind. It is not that animals are like us; we are like animals. Humans share 97.5 per cent of their genes with their nearest living non-human relatives, chimpanzees. Rabbits and pigs do not really wear clothes as they do in Beatrix Potter's, but their mammalian physiology is near enough to our own for us to think it worth while to use these animals for testing whether various substances may be harmful to ourselves.

Antibiotics, resistance to. *See* resistance.

Antibody, a protein, part of the >immune system in vertebrates. Antibodies are >immunoglobulins (abbreviation: Ig), of which there are several types). Each Ig molecule consists of four chains, each with a constant (C) region, a joining (J) region and a variable (V) region, the last occupying from a quarter to half of each chain's length. The function of an antibody is to recognize and combine with an invading >antigen. It is the variable region of the antibody molecule that is specific for the antigen and binds to it. Specificity is achieved not only by different sequences in the variable region, but also with different combinations of the variable regions between chains. Antibody molecules are produced by the >B lymphocytes in the spleen and lymph nodes, the appropriate cell line that makes a particular antibody being selected when the antigen specific to it enters the body (*see* clonal selection theory). *See also* monoclonal antibodies.

Anticoding strands. *See* sense strand.

Anticodon, the message part of a molecule of >transfer RNA. It matches on to the corresponding codon in messenger RNA (*see* protein synthesis). *See also* genetic code.

Antigen, any substance recognized by the body as foreign, against which >antibodies are made. Antigens include viruses, bacteria, fungi, parasites and non-living particles. *See* clonal selection theory; immune system.

Antiparallel alignment, the arrangement in which the two strands of DNA or RNA lie. They are each oriented from the 5′ to the 3′ end, and in antiparallel alignment the 5′ of one is opposite the 3′ of the other.

Antiserum, blood serum containing antibodies against a particular antigen. >Passive immunity is given to patients by injecting into them antiserum made by another animal so that the antibodies will enter their bloodstream immediately, rather than waiting for their own >immune system to produce them. A common example is administering or injecting antiserum against tetanus, the serum being derived from horses. Treatment with antiserum can produce an unpleasant side-effect (serum sickness) if the patient has an adverse reaction to other proteins present in the serum.

Apo-activator. *See* apo-inducer.

Apo-inducer, a protein that combines with the inducer to initiate gene transcription. *See* gene regulation.

Apomixis, reproduction without a sexual process, in plants. It can be either >vegetative propagation or >agamospermy, both resulting in groups of offspring that form a >clone. Sexual reproduction in plants is called >amphimixis.

Apo-repressor, a protein that becomes a repressor when combined with another molecule (co-repressor). *See* gene regulation.

Arboviruses, a group of viruses that can attack both insects and humans. Yellow fever and one type of encephalitis (inflammation of the brain) are caused by arboviruses. The genome is single-stranded RNA.

Aristotle (384–322 BC), Greek philosopher. Plato was his teacher for ten years, and he became tutor to Alexander the Great. He founded a school at the Lyceum in Athens, and most of his writings are notes for lectures that he delivered there. They cover an enormously wide range – all of the range of learning at that time. His system of logic was the basis for all the rest of his thought, and his views on politics and ethics have had an unparalleled influence on Western thought. So also have his thoughts on biology, which is less fortunate. Aristotle seems to have found biology, a practical subject with which his pupils would not have had to concern themselves, far less interesting than the important matters such as politics and virtue. He was an uncritical collector of biological 'facts', some quite bizarre, such as the notion that a female mouse can become pregnant before it is born. His *Historia animalium* (*c.* 335 BC) classifies animals into those that arise spontaneously from mud, etc. and those that reproduce sexually, the latter group being further divided into those that lay eggs and those that have live young. At least three of his ideas about genetics were totally wrong: *(1)* that some animals reproduce by >spontaneous generation (*see* p. 7); *(2)* that the father is responsible for the 'form' of the offspring (*see* p. 4);

and *(3)* that acquired characteristics can be inherited (*see* p. 9). It is fair to say that these were not Aristotle's own ideas; he was simply repeating the general beliefs and superstition of the time; once they had appeared in his writings, they acquired the prestige of his name and became virtually irremovable.

Artificial insemination, making a female pregnant by manually introducing sperm. It was first done to a spaniel bitch, by the Italian scientist >Spallanzani in the eighteenth century. It has been practised in humans since before World War II. However, donor insemination (formerly called AID, artificial insemination by donor, an acronym that was abandoned after the spread of AIDS), as distinct from artificial insemination by husband (AIH) has been extremely controversial on moral grounds. Jewish belief is against it, considering it to be 'bringing orphans into the world'. In Britain in 1948 the Archbishop of Canterbury stated that donor insemination should be a criminal offence, and a Royal Commission in 1960 reported that 'both society and the medical profession are opposed to AID [i.e. donor insemination]'. It was not until 1973, following the report by the Peel Committee, that donor insemination was officially approved as a treatment, and even then it was thought suitable to restrict its use as much as possible. Artificial insemination is widely used in cattle breeding. However, it is not permitted in racehorse breeding, in which the fee charged for the services of a stud horse is kept high because of the limit to the number of mares that he can mate during the season.

Artificial selection, selection deliberately carried out by humans, as opposed to natural selection. All examples of >domestication involve selectively breeding from organisms whose phenotype suits some purpose, and the changes so produced can be enormous. Artificial selection was understood by both Lamarck and Darwin, and the search for an equivalent force in natural conditions led the latter to his theory of natural selection. *See* selection limit.

Ascospore. *See* ascus.

Ascus, a sort of pod containing the four cells produced at >meiosis by certain fungi, e.g. >*Neurospora* and >*Saccharomyces* (yeast). Each of the four undergoes a mitosis division so that the ascus finally contains eight cells, known as ascospores. *See* tetrad analysis.

Asexual reproduction, reproduction by simple cell division. This commonly occurs in prokaryotes, and also fungi, protozoa and some plants. Bacteria normally reproduce asexually, but are also capable of a form of sexual reproduction known as >conjugation.

Aspergillus, a genus of fungi used in genetic experiments. The species most often used is *A. nidulans*. Another species, *A. fumigatus*, causes a lung disease, but the majority are among the many fungi found on decay-

ing food. *Aspergillus* is a eukaryote and may be either haploid or diploid (16 chromosomes).

Assortative mating, any system of mating in which mate choice is influenced by phenotype. In *positive assortative mating*, mates are more like each other than they would be by chance. This is quite commonly found, and has been noticed in human populations: in advanced countries, there is positive assortative mating both for height and for socioeconomic status (which traits are positively correlated with each other). In *negative assortative mating*, mates are less like each other than they would be by chance. From the point of view of the genetics of a population, positive assortative mating tends to increase homozygosity and negative assortative mating to decrease it.

Atavism, the reappearance of a form that used to occur in an ancestral stock, due to recessive alleles meeting each other or to recombination. A respectable term in modern ecology, it has a bad history in that it was used in the late nineteenth century to describe various human disorders as being reappearances of traits found in pre-humans (another example of the perversion of Darwin's theory of the descent of man). The popular term is 'throwback'.

Attenuation. There are two definitions of this word in genetics. First, it is the name given to a form of >gene regulation which operates when the gene product is an enzyme (or more usually a group of enzymes) that synthesizes an >amino acid. There is a leader sequence before the structural genes, and this includes a few repeats of the codon for the amino acid in question; if this sequence can be encoded – i.e. if there is already a sufficient quantity of that amino acid in the cell – the messenger RNA forms a terminator loop and transcription stops. If there is an insufficient amount of the amino acid – that is, not enough molecules of its transfer RNA – transcription of the repeat codons cannot take place, the terminator loop is not formed, and the gene(s) are transcribed through to the end. Therefore shortage of the amino acid turns the gene on, and when the supply is sufficient it is turned off.

Second, attentuation is the word used for loss of virulence in a strain of viruses or bacteria. Viruses can be attenuated by treating them with ultraviolet radiation; attenuated strains of bacteria are obtained by selection *in vitro*. Both types of attenuated micro-organisms are used in >vaccination.

Attenuator, a sequence in a messenger RNA molecule where transcription stops, in the system where the gene is regulated by >attenuation. It comes immediately before the coding sequences.

Australopithecus. *See* human evolution.

Autogamy (from the Greek *auto*, 'self', and *gamos*, 'marriage'), self-fertilization. Usually the term refers to self-fertilization in plants, but it

can also describe the process in which a >*Paramecium* divides its macronucleus to form a new micronucleus and macronucleus after cell division.

Auto-immune diseases, a group of diseases in which the symptoms are caused by the >immune system attacking the body's own tissues. In many of these diseases there seems to be a genetic element, in that the disease occurs among relatives; an example is systemic lupus erythematosus, in which many different organs, including the skin, heart, spleen and kidneys, are attacked as 'foreign'. A few auto-immune diseases have been linked to specific types in the >HLA complex, part of the genetic system controlling the immune response. >Rheumatoid arthritis is known to be four times more frequent in people with HLA type DR4, and >ankylosing spondylitis is almost entirely confined to people with HLA type B27. None the less, it is not possible to say that merely having this or that HLA type is enough to cause an auto-immune disease; there must be some trigger factor, such as infection by a virus, which is more dangerous to one HLA type or another.

Autogenous regulation. *See* autoregulation.

Autopolyploid, a >polyploid organism in which all the sets of chromosomes come from the same species. *See* allopolyploid.

Autoregulation, the system in which a gene's activity is controlled by its own gene product. *See* gene regulation.

Autosexing, a technique for identifying the sex of organisms before the age at which the sexual organs are differentiated. The classic example is chicks: in certain strains the males and females are a different colour, distinguishable at one day old.

Autosome, a chromosome that is not a sex chromosome. The autosomes always outnumber the sex chromosomes (of which there are only two).

Avery, Oswald Theodore (1877–1955), American doctor and bacteriologist, who discovered that the genetic material in chromosomes is >DNA. He repeated F. Griffith's experiment in which non-virulent viruses are transformed into virulent ones by something contained in an extract of heat-killed virulent viruses. In 1944 Avery showed that the transforming factor was DNA, a molecule that no biologist was very interested in. Avery did not make the claim that genes consisted of nothing but DNA, but he did state that DNA was an important element of genetic material. This finding made very little impact at the time.

B

B chromosome, one that is an extra to the normal diploid set (the A chromosomes). B chromosomes are more often found in plants than in animals (though grasshoppers specialize in them). Because they do not seem to have any genetic effect it has been suggested that they are 'selfish chromosomes' (*see* selfish DNA).

B lymphocyte, a type of white blood cell, part of the immune system. The B lymphocytes' function is to produce the antibody molecules that bind antigens in humoral immunity (*see* immune system). When an antigen arrives in the body, it is recognized by a virgin B lymphocyte in the bone marrow (*see* clonal selection theory); this cell is stimulated by the antigen into proliferating and establishing a clone of descendant cells (plasma cells) in the spleen and lymph nodes which produce large amounts of antibodies. B lymphocytes (so called because they mature in and are released from the bone marrow) recognize antigens by their three-dimensional shape (unlike T lymphocytes).

B_1, B_2, etc., designations for the generations involved in a >backcross. The cross between the first hybrid generation (>F_1) and the parental type is the first backcross, and the offspring of this are the B_1 generation. If individuals from the B_1 generation are again backcrossed to the parental type, the resulting offspring are the B_2 generation, and so on. B_2 is also sometimes used to designate the offspring of the backcross of F_1 to the other of the two original parental types.

Back mutation, a >mutation that happens to an already mutant allele, correcting the mutation and restoring the >wild type. It may be an exact reversal of the first mutation, or it may really be a second >forward mutation that happens to have the effect of cancelling out the effect of the first.

Backcross, a cross between >heterozygous individuals of the >F_1 generation and individuals of one of the parental genotypes. The result is a generation in which the hybrid and parental types are present in the ratio of 1:1 – for example, *Aa* (the F_1) × *AA* (the parental type) gives ½ *AA* and ½ *Aa*. The term 'backcross' is also used for the cross between the F_2 generation and one of the parental types. A >testcross is a particularly useful form of backcross.

Background genotype, the part of the genotype that is concerned with traits not normally discernible in the >phenotype.

Bacteria, a large group of single-celled organisms. They can occur either free-living in air, water, soil, etc., or they may be parasitic upon a host plant or animal. Only a minority cause any disease, and more are actually useful to their hosts, e.g. in breaking down food as part of the digestive process (*see Escherichia coli*). The three basic shapes are *bacillus* (rod), *spirillum* (spiral) and *coccus* (spherical), but some bacteria may link up to form chains or clumps. Much genetical research is done on bacteria because, being >prokaryotes, they have simple genetic machinery. Bacteria normally reproduce asexually (i.e. by simple binary cell division), but occasionally 'conjugate' to reproduce sexually (as discovered by Joshua >Lederberg and E. L. >Tatum). The word itself comes from the Greek for 'walking stick'.

Bacteriophage, a virus that infects bacteria. A 'phage' (as it is usually called) consists of an outer capsule made of protein and a core of nucleic acid, which can be either DNA or RNA. The phage attaches itself to the cell wall of the bacterium, sometimes by means of a spiky 'tail', and delivers its DNA or RNA into the host cell. The host then replicates the virus, either by treating the DNA as one of its own genes or by using the RNA as mRNA; in either case, the phage's entire genetic constitution is expressed by the host cell's protein synthesis apparatus. Thus the host cell begins to contain a number of viruses that the cell itself has made (this process was discovered by >Hershey). Sometimes the mature viruses leave the host cell by being extruded through the cell wall, but more usually the result is >lysis, when the viruses produce an enzyme that bursts the cell wall. Phages (especially the >lambda and >T4 varieties) have been intensively studied because the way they get bacteria to utilize their DNA may be comparable to the way in which genes are switched on in normal cell metabolism (*see* gene regulation).

Von Baer's law, the rule that, in their earlier stages of development, organisms resemble each other more than they do at maturity. It was formulated by the Estonian embryologist Karl von Baer (1792–1876) who also discovered the mammalian ovum, and published in his *Über der*

Entwickelungs Geschichte der Thiere (part I 1828, part II 1837, *History of the Development of Animals*). *See also* Haeckel's law.

Balanced lethal, a system in which a population is consistently heterozygous for two (or more) recessive >mutations that are lethal when homozygous and are not linked. Let us take a two-locus example. All individuals have genotype *m*+ *n*+ (where *m* and *n* are the mutant alleles and + are the wild-type alleles). This is a >dihybrid cross, so the result is a 9:3:3:1 ratio, in which the first nine are all viable because they are double heterozygotes, but the remaining seven are non-viable, being homozygotes for one or other of the lethal recessives. The survivors appear to breed true, but only in the sense that all their *surviving* descendants are genotypically the same as themselves; they have had other offspring that have died, so this is not the same thing as >true breeding.

Balbiani rings. *See* puff.

Baldness, a classic example of a sex-limited trait. Premature balding only occurs in men, but this does not mean that the gene for it is on the Y chromosome. It can be shown in a pedigree that it is an autosomal dominant, which can be transmitted to either daughters or sons. A female who has the allele will not herself go bald but will pass on the baldness allele to her offspring; thus half her sons will go bald. The gene is only expressed in men, because it requires the presence of testosterone (*see* sex hormones).

Baltimore, David (1938–), American molecular biologist; professor of microbiology at the Massachusetts Institute of Technology 1968–90; president of Rockefeller University since 1990; discoverer of >reverse transcriptase. After important work on how the polio virus replicates, Baltimore found that in some tumour viruses there is an enzyme that can make a DNA molecule by reading off from RNA. This the exact opposite of what the central dogma states, i.e. DNA → RNA → protein. Baltimore's discovery of reverse transcriptase won him the 1975 Nobel prize for physiology or medicine (jointly with H. Temin who had independently made the same discovery).

Banana. *See* triploidy.

Banding, the pattern of light and dark regions in chromosomes revealed by various staining techniques. The bands can be used to identify individual chromosomes within the karyotype. In the giant >polytene chromosomes in *Drosophila* it is believed that the light interbands mark the areas where >housekeeping genes are located and the dark bands mark the >luxury genes. The different appearance is due to the fact that where genes are being transcribed the DNA has to be loosely wound, whereas the DNA for inactive genes is highly condensed and shows up darker (*see* heterochromatin).

Barr body, a dark-staining particle seen in the nucleus of any cell (when it is not dividing) of a female animal. The Barr body is actually one of the two X chromosomes that has become inactivated (*see* Lyon hypothesis). As the Barr body can be seen in any non-dividing cell, it is extremely easy to distinguish the sex of an individual from a small sample (e.g. in prenatal sex testing), though the Barr body cannot be seen in red blood cells because they have no nuclei. A Barr body is also seen in the cells of an individual whose complement of sex chromosome is XXY (*see* Klinefelter's syndrome) even though they are phenotypically male, but an individual with >Turner's syndrome (sex chromosomes XO), phenotypically female, has no Barr body. A person with three X chromosomes will have two Barr bodies in each cell, as there is always one Barr body fewer than the number of X chromosomes. In insects and others where it is the male that is the >homogametic sex Barr bodies are seen in the cells of males. *See also* heterochromatin.

Barrier, in genetic terms, any impediment to the exchange of genes between one population and another. Barriers are usually geographical features, but might be other factors, e.g. different behaviour patterns. *See also* isolating mechanism.

Base-pair (abbreviation: 'bp'), one of the pairs of nitrogenous bases that form the core part of a molecule of >DNA or >RNA. They are either >purines (guanine and adenine) or >pyrimidines (cytosine, thymine and uracil), and in each pair there is one of each, guanine always pairing with cytosine, and adenosine always pairing with thymine (in DNA) or uracil (in RNA). Genes are often measured in terms of how many base-pairs they consist of (*see* kilobase).

Bates, Henry Walter (1825–92), British naturalist. Accompanied by >Wallace, Bates went to South America where between 1848 and 1859 he collected a vast number of specimens, including some 8000 species (mostly insects) that had never before been described. His theory of >mimicry was one of the pieces of evidence that Darwin used to support his idea of evolution through natural selection.

Bateson, William (1861–1926), British geneticist; professor of biology at Cambridge and director of the John Innes Institute; inventor of the term 'genetics'. During the 1890s Bateson's interest in evolution led him to the conclusion that species evolve by 'jumps' rather than gradually, and in trying to prove this experimentally he came up with ideas very similar to Mendel's. When Mendel's work was rediscovered in 1900, Bateson translated it into English and became its leading advocate. He found the first Mendelian genes in animals, and correctly interpreted Garrod's findings as showing Mendelian genes in humans. He obtained some results that did not match Mendel's law of independent assortment, and this led to T. H. Morgan's discovery of linkage.

Bayes' theorem, a way of calculating the relative probability of an event. It is much used in medical genetics (*see* genetic counselling) to predict the probability of a person being a carrier of a deleterious recessive gene in a family where such a gene is known to be present. It combines a prior probability, based on the occurrence of the gene in the pedigree, with a conditional probability, based on the phenotype of the individual, which together lead to the posterior (or final) probability.

Beadle, George Wells (1903–), American geneticist; professor of biology at the California Institute of Technology for 14 years before going to the University of Chicago in 1961. With his microbiologist colleague Edward >Tatum, in the 1940s Beadle investigated mutant genes in the pink bread fungus >*Neurospora crassa*, a good subject for genetical research since it is haploid (i.e. has only one set of chromosomes) so that recessive mutants are expressed. Beadle and Tatum gave the mutant strains of bread fungus different mixtures of nutrients to grow on, and found that specific mutations led to an inability to digest one nutrient. Their conclusion was that each mutation caused the loss of a particular enzyme, hence the statement 'one gene, one enzyme'. They did not know that this idea had been put forward by Garrod some 40 years earlier. Beadle and Tatum shared the 1958 Nobel prize for physiology or medicine with Joshua >Lederberg.

Beneden, Edouard van (1846–1910), Belgian zoologist. Beneden, a pioneer in the study of chromosomes, made the discovery that the number of chromosomes is the same in all the cells of an animal's body, except for those found in the sex cells, which have half. He also discovered that the chromosome number is not the same for every species. Beneden did not publish an adequate interpretation of his findings; but they were taken up by others, notably Boveri, de Vries and Weismann.

Berg, Paul (1926–), American geneticist; professor of biochemistry at Stanford University since 1959. Berg took up Francis >Crick's 1955 suggestion that there must be an intermediate molecule between the RNA template and the protein being formed during protein synthesis; within a year, he had found the first example of tRNA, or transfer RNA, which is this intermediate. Berg was also a pioneer of techniques for introducing specific genes into host bacteria (*see* recombinant DNA), an important part of genetic engineering. He was one of the first to see the risks involved in recombinant DNA research, and urged extreme caution in pursuing it. He shared the Nobel prize for physiology or medicine in 1980.

Bidirectional replication, DNA replication in which, from a single starting point, replication proceeds in both directions along the molecule.

Biffen, Sir Rowland Harry (1874–1949), British geneticist. Biffen was among the first to put Mendel's laws to practical use when they were rediscovered in 1900. He was then teaching agricultural botany at Cambridge University, and he at once showed that physiological traits, as well as the visible ones that Mendel had studied, could be controlled by single genes. He was therefore able to breed strains with characteristics such as resistance to fungal infection. His wheat 'Little Joss' was the outstanding variety until after World War II.

Bimodal distribution, a statistical distribution in which the individuals cluster round two different values. On a graph, this would show as a curve with two peaks separated by a dip. Continuous variation due to a single pair of alleles and with environmental influence shows a bimodal pattern, if there is complete dominance.

Binary character, a genetically determined trait that exists in two states, which can be recorded as + or −, 0 or 1, etc. An example is the presence or absence of horns (in some breeds of cattle and sheep; some are more complicated, having multiple horns). *See also* multistate character.

Binomial nomenclature, the system, devised by >Linnaeus in the eighteenth century, under which all species have been given a two-part name. The first part of the name is the genus and is a noun; the second is the adjective that designates the particular species (*see* p. 13). If the species is the type-species of the genus, the two parts may be the same, e.g. *Gorilla gorilla*. Both parts of the name are in Latin (which is useful in that it is to some extent still a *lingua franca*), and printed in italic. The specific name always begins with a small letter, even when it is based on a proper name, e.g. *Fregata andrewsi* (one of the frigate birds). *See also* classification.

Biological warfare, the use of pathogens (disease-causing agents such as viruses, bacteria, fungi, parasites) as weapons. The target may be humans, or it may be the enemy's agricultural animals and plants. Biological warfare was banned under the 1925 Geneva Convention, but only for 'first strike' use; this proviso enabled governments to develop biological weapons as long as they could be described as 'defensive'. The 1975 Biological Weapon Convention (signed by over 100 countries) goes further, banning all development and production of biological weaponry; unfortunately it allows research for defence and does not define the amount of stock that can be held for research as distinct from operational purposes. It is believed that at least ten countries (including Britain and the United States) have developed a biological warfare technology; the United States spends $60 million per year on its research programme. This research probably consists to a large extent of genetic engineering, manipulating the virulence of common pathogens, and also of immunization procedures for one's own troops or population. Deliberate ecological disturbance, e.g. the use of defoliants, is in a sense biological warfare too.

Biometrics, the science of applying statistical methods to biological problems. It is much used in plant and animal breeding for agriculture, where the data typically is concerned with >continuous variation.

Biotechnology. In its narrowest sense, this term means the use of micro-organisms to make an end-product. This includes such things as the ancient technique of using yeast to make wine from grape juice, as well as such recent ones as the production of quantities of once-scarce human hormones from genes >cloned into >*Escherichia coli*. (*See* genetic engineering.) Biotechnology can also mean anything that improves human utilization of animals and plants, such as >artificial insemination and >embryo transfer for cattle, or the improvement of crops by selection or genetic manipulation. Bioengineering is the word used for the development of such things as artificial limbs, kidney machines and so on. Biotechnology can make an immense contribution to 'green' environmental policies. One key example is the production of fuel for road vehicles by using yeasts to make alcohol from vegetable matter; this means that vehicle fuel can be a renewable resource, in contrast to the finite stocks of oil and gas that are fast running out (of course, vehicles themselves would have to be changed). Another example is the use of ecological manipulation (e.g. the introduction of sterility genes into wild populations of insects) rather than chemical pesticides to protect crops.

Bivalent, a pair of homologous chromosomes at the time when they lie alongside each other during the first meiotic division (*see* meiosis). Each chromosome in a bivalent consists of two >chromatids.

Blastocyst, a stage in the >embryogenesis of mammals, prior to implantation. It is the first stage at which there is any differentiation among the ball of cells, which then number about 64. The blastocyst consists of an outer layer of protective cells (the *trophectoderm*) surrounding a fluid-filled cavity containing the inner cell mass that goes on to develop as the embryo. The human blastocyst is about 0.15 mm in diameter.

Blastomere, one of the undifferentiated cells in the very early life of the embryo. It makes up the >blastula. *See* embryogenesis.

Blastula, a stage in the >embryogenesis of animals that consists of a ball of cells which have not yet undergone any differentiation. It is a hollow ball of cells (called blastomeres) either one or several layers thick. The next stage is the >blastocyst.

Blending inheritance, the mistaken idea that heredity involves an irrevocable mixing, as of fluids, between the two parental contributions. *See* pp. 7–8.

Blindness, total or partial loss of sight. It is usually caused by disease or accident, but there are a few forms of blindness that are genetic. One is early-onset blindness, which starts by the age of five and affects one child in 10,000; it is inherited as an autosomal recessive. Another form is caused by a dominant gene, and affects about the same number of children, though the blindness does not begin until they are about ten. The inherited cancer >retinoblastoma also causes blindness, and other genetic diseases have blindness as a secondary effect. *See also* retinitis pigmentosa.

Blood group systems, the various sets of >antigens found on human red blood cells. Over 30 blood group systems have been discovered since the first – the >ABO – in 1900. Only the ABO and the >Rhesus systems are normally of medical importance. *See also* Duffy blood group system; MN blood group system.

Blunt end, the end of a DNA molecule where there are two strands of DNA right down to the last base-pair, as opposed to one where the last few bases on one side are unpaired (*see* sticky end). *See also* ligation.

Bombyx mori. The silk moth. *See* silk.

Bottleneck, in genetic terms, a decrease in population size that suddenly reduces the amount of genetic diversity in the population, especially if the survivors are not genetically a random sample of the former population.

Boveri, Theodor Heinrich (1862–1915), German zoologist. Boveri started from Beneden's observations that the chromosome number is constant in any species and that a sperm and an egg each contain a half-set of chromosomes that unite at fertilization to give the complete number found in all other cells. He went on to observe (in sea urchins) that an individual that has a deficient number of chromosomes will develop abnormally; this was an added piece of evidence that chromosomes are indeed the 'packages' in which the hereditary material is carried.

Boyer, Herbert Wayne (1936–), American biochemist; professor of biochemistry at the University of California at San Francisco since 1976, and co-founder of Genentech, one of the earliest commercial genetic engineering firms. In 1973 Boyer developed the technique of splicing together DNA from two different plasmids in >*Esherichia coli* to produce a hybrid functional 'gene', known as a >chimaera.

Bp, abbreviation for >base-pair, used as a unit of length to describe a sequence of DNA.

Breakthrough, an individual that is >homozygous for a lethal recessive mutant gene but none the less manages to survive (though not necessarily to reproduce).

Brenner, Sydney (1927–), South African-born British molecular biologist; director of the Medical Research Council Laboratory of Molecular Biology 1979–86, and since 1986, director of the Molecular Genetics Unit at the same site; joint discoverer of the genetic code. In 1957 Brenner joined the MRC's molecular biology laboratory at Cambridge, where the structure of >DNA had been discovered by Francis >Crick and James >Watson four years earlier. His work showed that the bases in DNA form a code if read in threes, and that these >triplets (or >codons) are not overlapping; GCATAGATC is GCA–TAG–ATC, not GCA–CAT–ATA and so on. Brenner and Crick found in 1961 that the molecule that carries the code from the DNA in the chromosome to the >ribosomes is messenger RNA (mRNA). Brenner was also the prime mover of the British project to map the human genome (>Human Genome Project and pp. 36–8).

Buffon, Georges Louis Leclerc, Comte de (1707–88), French scientist, one of the first to suggest the evolution of species. He was keeper of the Jardin du Roi (now Jardin des Plantes) in Paris, and of the French royal museum of natural history; his 44-volume catalogue of the latter (*Histoire naturelle*, 1749–1804) grew into the first serious attempt to describe the whole natural world. On the evidence of certain vestigial features such as the side toes on pigs, Buffon suggested that similar animals are descended from 'common ancestors'. This idea, and Buffon's realization that species in nature are variable not uniform, were influences upon Darwin.

Burkitt's lymphoma, a tumour of the lymph glands, common in some parts of West Africa. It appears to be triggered by a virus infection (*see* p. 34), and is one of the commonest cancers found in people with AIDS.

Burt, Sir Cyril Lodowic (1883–1971), British psychologist, remembered for his falsification of results in studies on the heritability of intelligence (*see* IQ). Burt was professor of psychology at University College, London, and a passionate believer in selective education; his advice to the British government at the time of the Education Act 1944 was crucial in setting up the two-tiered system of grammar and secondary modern schools. Burt claimed to have data derived from studies of identical twins separated at birth that proved intelligence to be almost entirely genetically determined; but in 1976, five years after Burt's death, the *Sunday Times* journalist Oliver Gillies exposed the data as fraudulent. It seems that Burt did originally have some rather unsystematic data on a few sets of twins but that this information had been lost in the war. Therefore, when he came to write his most polemical papers, after his retirement, he made up data to suit his argument, and invented a research assistant who never existed. For some of his data, *see* twin studies.

C

C region, or constant region, part of the heavy and light chains of an >immunoglobulin molecule. *See* antibody.

***C* value**, the amount of DNA in a genome (in its haploid condition). Very generally, the *C* value is higher in organisms that are evolutionarily more advanced, but sometimes closely related species have widely different *C* values, and there are sporadic cases of very high *C* values with no obvious explanation (this is known as the *C*-value paradox); for example, the *C* value of humans is 3.0 picograms per cell, of yeast 0.024, of broad beans 24.0, of geese 1.4. In any case, there seems to be far more DNA per cell in higher animals than can be necessary for the genes in that animal; much must be redundant or repetitious.

Camerarius, Rudolf Jakob (1665–1721), German botanist; simultaneously professor of medicine and director of the botanic garden at Tübingen; discoverer of sexual reproduction in plants. Camerarius was not the first to notice that pollen causes fertilization in plants (it was known, in some plants at least, to the ancients; *see* Theophrastus), but in 1694 he proved experimentally that the male and female flowers of plants are exactly analogous to the sexual parts of animals, with stamens the male organs and pistils the female.

Canalization, the hypothetical tendency of an organism during development (particularly >embryogenesis) to follow the right pathways, irrespective of any adverse influence from the environment.

Cancer, a blanket term for a group of diseases with the common factor that they involve the uncontrolled proliferation of cells. Humans, animals, birds and plants can all be afflicted with cancer. Virtually every part of the body is susceptible, though some are much more commonly affected than others (cancers of the lung, stomach, bowel, breast, and prostate are among the frequent ones in European populations; incidences vary enormously between populations). The cancer sometimes remains localized in one tissue, and sometimes spreads via the bloodstream or lymphatic system to other parts of the body (metastasis). The many different cancers have different causes, and only in a few can a single cause be identified. There is an environmental factor is some – for example, the high frequency of cancer of the scrotum in chimney sweeps was due to their contact with soot, and radiation and chemical pollution are today strongly implicated in some cancers. Others are caused by a virus, for instance, Burkitt's lymphoma (*see* p. 34). Genetic analysis of the family histories of cancer patients shows that, while most of the common cancers show no genetic pattern, a few particular cancers do run in families.

Some cancers are caused by a recessive gene. One of these is >retinoblastoma, tumour of the eye. Children who are homozygous for the abnormal allele develop eye tumours early in life. In other cases, there is a later onset, and it seems that these patients are heterozygotes in whom the protection given by the normal allele is lost when the allele itself is lost in one cell (possibly by the deletion of a small segment of a chromosome or non-disjunction during cell division). Other cancers have an inheritance pattern that suggests a dominant allele, but with variable penetrance (i.e. not everyone who carries the gene will get the disease); an example is the type of breast cancer that develops in both breasts before the menopause. There are also cancer-related conditions that are inherited; one of these is >polyposis coli, which is often the precursor of cancer of the colon and is inherited as an autosomal dominant. One of the most puzzling findings is of 'cancer families', in which the members of the family get different kinds of cancer, but the getting or not getting of the disease is inherited as a single gene; these families are very rare.

Cap site, a modification to the 5′ end (the 'front') of a messenger RNA molecule in eukaryotes. It seems to assist the molecule in attaching itself to a ribosome. The other end has a >poly-A tail.

Carcinogen, any environmental agent that causes cancer. A carcinogen may be a chemical substance, such as tar (associated with lung cancer) or aniline dyes (bladder cancer), or it may be a form of radiation, such as >ultraviolet light (skin cancer) or the radiation leakage from nuclear power stations that is now suspected of causing leukaemia in the children of male power-station workers.

Carrier, an individual who is heterozygous for a recessive gene. Such individuals do not show any of the effects of the gene but transmit it to their offspring. The term is also used for a person who has become infected with a bacterium or virus, and although he or she does not have the symptoms of the disease in question, he or she can pass it on to others. *See also* X linkage.

Carrying capacity, the maximum number of organisms that can inhabit a given area. *See* density-dependent factors.

Caste, in ecology, a group of individuals within a population who share some morphological or behavioural characteristic. It is a useful term for describing the different groups within a colony of social insects. In human society, especially in the Indian subcontinent, castes are hierarchical, closed social units, each with prescribed occupations and with obligatory endogamous marriage. Although there is some gene flow between castes, endogamy has been strictly enforced for long enough for there to be differences in gene frequencies.

Catalyst. *See* enzyme.

Catastrophism, the idea that the geological features of the Earth were formed in a series of violent and catastrophic events. Baron Cuvier (*see* p. 14) believed that the succession of fossils was accounted for by a series of catastrophes, after each of which life was created anew. It was essentially an anti-evolutionary point of view, and, as far as geology went, was replaced by the theory of uniformitarianism advanced by the British geologist James Hutton in the late eighteenth century (*see* pp. 12–13). Catastrophism is now taken to mean that the evolutionary process does not go at a uniform speed but is fairly slow until enormously speeded up in the aftermath of catastrophes. *See also* gradualism.

cDNA. *See* complementary DNA.

Cell cycle, the periodic phases in a cell's life. The two major phases are >mitosis (cell division) and >interphase, the period between divisions when the cell is growing and then doing whatever it is specialized to do. Interphase is itself subdivided into three parts: G_1 (gap 1), in which each chromosome is represented by only one >chromatid; S (synthesis), during which DNA is replicated so that each chromosome again has two chromatids; and G_2 (gap 2), leading to mitosis (also called 'M phase'). Interphase takes up about 90 per cent of the time of a >eukaryotic cell, with the duration of G_1 being anything from a few hours to several years long.

Cell line, a population of cells maintained in tissue culture. If the original cells were differentiated, i.e. had become specialized for some function, the cells will continue to show that differentiation when kept >*in vitro*. Over a long period of time, cumulative chromosome changes will become apparent and the cell line may finally become unviable. *See* HeLa cells.

centimorgan, a unit (named in honour of T. H. >Morgan) used to measure the frequency of >crossing-over and therefore the distance apart of linked genes: 1 centimorgan is equal to a 1 per cent frequency of cross-over, and is therefore equivalent to a >map unit. The original unit, the morgan, is equal to 100 per cent cross-over.

Central dogma, the basic principles of molecular biology as stated by Francis >Crick. They are:

- DNA is replicated to produce more DNA.

$$\text{DNA} \rightarrow [\text{replication}] \rightarrow \begin{cases} \text{DNA} \\ \text{DNA} \end{cases}$$

- DNA is transcribed into RNA which is translated into protein.

$$\text{DNA} \rightarrow [\text{transcription}] \rightarrow \text{RNA} \rightarrow [\text{translation}] \rightarrow \text{protein}$$

Common to both parts is the idea that DNA is the *source* of information, not the recipient. However, an exception happens with >reverse transcriptase, an enzyme that can make a DNA molecule from an RNA template. (*See also* pp. 30–1.)

Centriole, one of a pair of organelles in the cell nucleus, which make the fibres for the >spindle and also form the poles from which the spindle fibres radiate in >anaphase.

Centromere, the point at which the two arms of a chromosome join together end to end. The centromere is also the point at which the chromosome is attached to the spindle fibres during cell division (*see* meiosis; mitosis). Chromosomes may have the centromere right at the end (>telocentric), at or near the middle (>metacentric) or towards one end (>acrocentric). The DNA at the centromere consists of repeated short sequences that carry no genetic information. *See also* acentric chromosome.

Chargaff, Erwin (1905–), Czechoslovakian-born American biochemist; professor of biochemistry at Columbia University, New York, from 1952 until his retirement; famous for his basic work on the properties of DNA. Working at a time (the late 1940s) when few biologists were interested in the nucleic acids, Chargaff discovered two very important facts about DNA. The first was that each species has its own characteristic DNA, differing from other species in the relative amounts of four nitrogenous bases: adenine (A), cytosine (C), guanine (G) and thymine (T). The second was that, in any sample of DNA, the amount of A is equal to the amount of T, and C is equal to G. These two principles became known as 'Chargaff's rules', and were used by Francis >Crick and James >Watson in their search for the structure of DNA; the significance of the rules was immediately apparent once that structure was established.

Chiasma (pl. chiasmata), the point at which non-sister chromatids of homologous chromosomes join during >crossing-over. As seen microscopically, each chiasma does indicate a crossing-over event in genetic terms, but the site of the chiasma does not correspond exactly to the molecular point of exchange.

Chimaera (or chimera; from the name of a creature in Greek mythology, with a lion's head, goat's body and snake's tail), an organism which has cells of two or more different genotypes, derived from different zygotes. One example is the >aggregation chimaera of two different strains of mice. A chimaera created by adding a single genetically marked cell to an 8-cell embryo can be used to study embryonic development (*see* fate map). Chimaeras can also be created out of two different species, e.g. sheep and goat. A plant chimaera is a plant in which part of one individual has been grafted on to another and the resulting plant contains cell lines derived from both; a flowering plant could thus bear flowers of two different strains or species. *See also* mosaic.

Chi-square test, a statistical technique that tests whether or not the observed occurrence of something fits with its expected occurrence. It can only be used with either/or variables, not continuous ones. An example would be the number of patients who died or who survived under *(a)* the traditional treatment or *(b)* treatment with a new wonder drug. Testing the >null hypothesis – in this case, the proposition that the new drug is no improvement – one compares the proportion of survivors to see whether it is or is not significantly different from what was expected (i.e. from what it had been with traditional treatment).

Chlamydomonas reinhardtii, a single-celled green alga, often used in genetical research. For most of its life cycle it is haploid, only becoming diploid when two individuals fuse in order to reproduce; the resulting >zygote immediately divides by meiosis to form four new individuals.

Chlorella, a genus of green alga. It is studied for the genetic control of photosynthesis.

Cholesterol. *See* hypercholesterolaemia.

Chorionic villus sampling (CVS), a technique for sampling the genetic make-up of the foetus, using cells from the chorion, a layer derived from part of the blastocyst (*see* embryogenesis) and therefore genetically the same as the foetus but attached to the wall of the uterus. Enough of these cells can be gathered for their DNA to be directly analysed, or else the cells can be cultured for the chromosomes to be visually examined. The procedure can be carried out at 8–10 weeks of pregnancy and involves much less stress for the mother than amniocentesis, as it is done with a catheter through the cervix. There is probably a higher risk to the foetus than with amniocentesis (of the order of 3.5 per cent casualties) but this is disputed because there is a significant loss through spontaneous abortion at about that time of a pregnancy. *See* prenatal diagnosis; polymerase chain reaction.

Chromatid, half of a duplicated chromosome. When a chromosome is about to divide, it doubles itself longitudinally into two sister chromatids joined at the >centromere; each chromatid is therefore equivalent in content to a single chromosome. At cell division (*see* meiosis; mitosis), the chromatids part and each then becomes a new chromosome.

Chromatin, the substance of which chromosomes are made. It consists of >DNA and the protein >histone in roughly equal proportions, arranged in bead-like units called >nucleosomes. It gets its name from the Greek *chroma*, 'colour', because it appears coloured after biological staining. Regions of chromosomes that are not genetically active have chromatin that is highly condensed (*see* heterochromatin), whereas active regions have a much more open configuration (*see* euchromatin).

Chromomere, a very small concentrated region of >heterochromatin within a chromosome, seen only during meiosis. The chromomeres are landmarks by which different chromosomes can be identified. In >lamp-brush chromosomes they show up as loops, and in >polytene chromosomes they look like bands.

Chromosome, the structure which carries the genes. The chromosomes lie within the cell nucleus, and divide when the cell divides: this process (called >meiosis) is the key to heredity. It is useful to think of a chromosome as a string of beads, each gene being a bead (and its place on the string being its >locus). The genes are made of >DNA, and the chromosome also contains a protein called >histone; this mixture is known as >chromatin. The chromosomes are not visible within the cell (because they unwind themselves into a structure too loose to be resolved by a microscope) except when cell division is about to take place. Then they become condensed, and it can be seen that they consist of two arms attached end-to-end at a point called the >centromere. At this stage the chromosomes also split themselves into two identical strands called >chromatids (*see* mitosis). The complete set of chromosomes is called the >karyotype. This includes the two >sex chromosomes plus the general-purpose chromosomes known as the >autosomes. The number of chromosomes varies between species from three pairs to over a hundred (*see* p. 23), but bacteria each have only a single chromosome in the form of a ring.

Chromosome abnormalities, any irregularity in the number or gross structure of the chromosomes. The great majority of these abnormalities are lethal, the organism being unable to sustain the loss of so much genetic material or the surplus genetic dose, as the case may be. It is estimated that about 10 per cent of human conceptions are spontaneously aborted (miscarried) because of chromosome abnormalities.

Over 60 disorders in humans caused by chromosome abnormalities are known, and the most common involve the sex chromosomes. A female may have either only one X chromosome (*see* Turner's syndrome) or three, instead of two; a male may be XXY (an extra X; *see* Klinefelter's syndrome) or XYY (*see* XYY syndrome), instead of XY. The overall incidence of the sex-chromosome abnormalities is about 2 per 1000 live births. Abnormalities involving one extra autosome (i.e. >trisomy) are also quite common, totalling about 1.7 per 1000 live births, and they include the well-known trisomy 21, commonly known as >Down's syndrome. There are also abnormalities in which a chromosome is missing (*see* monosomy). All these arise from non-disjunction of the chromosomes when ova and sperm are being made. Structural abnormalities such as major translocations and inversions occur in about 1.3 per 1000 live births; they may be the result of radiation damage to chromosomes, but little is known about what causes them in normal circumstances.

***Cis* configuration**, in an individual that is heterozygous one of the two possible arrangements for the alleles for mutations at two linked loci. *Cis* (from the Latin, 'on this side of') means that the two mutant alleles are both on one chromosome and the two wild-type alleles are on the homologous chromosome. This is also known as 'coupling'. The opposite is >*trans*.

Cistron, a sequence of DNA that codes for a >polypeptide, i.e. that functions as a gene. *See* intron.

Clade, a group of species or other taxa that share a common ancestry and are more closely related to each other than to any other taxa. *See* phylogenetic classification.

Cladistic distance, a measure of how closely two species (or other >taxa) are to each other, in terms of the number of branching points between them in the tree-diagram of their ancestry. *See also* phenetic distance.

Cladogenic evolution, evolutionary change in which there is a repeated pattern of one species splitting off from another. The resulting 'tree' (*see* dendrogram) has many branches, with numerous species co-existing at the same time. An example is the natural populations of the genus *Drosophila*, with nearly 1500 species known. *See also* phyletic evolution.

Classification, the systematic ordering of all living things. There have been many different attempts at this (*see* Aristotle for an early one), and the version in use today is doubtless not definitive. For a long time it was accepted that there were two kingdoms, Animalia (animals) and Plantae (plants), but the rise of microbiology has caused a rethink about the simpler life forms, so that the usual scheme now is to have five kingdoms:
(1) Monera: all the >prokaryotes, including >bacteria and blue-green algae.
(2) Protoctista: single-celled >eukaryotes, including amoebae, higher algae, slime moulds, Protozoa, etc.
(3) Fungi: fungi of all sorts from yeast (>*Saccharomyces*) to mushroom-like forms. The genetically important >*Aspergillus* and >*Neurospora* are in this kingdom.

(4) Plantae: all plants from the simplest (mosses and liverworts) to the trees and >flowering plants.
(5) Animalia: including sponges, sea anemones, worms, molluscs, spiders, millipedes, insects, fish, amphibians, reptiles, birds and mammals.
Note that viruses are not included, since although they are made of the same stuff as living things, they are not capable of reproducing themselves except via a host organism. *See also* origin of life; taxonomy.

Cleavage, the division of the fertilized egg into, first, a two-celled stage, then the four-celled, etc. (*see* embryogenesis). The word is also used for the cutting of a DNA molecule by an >endonuclease.

Cleft palate, a condition associated with more than one congenital disorder. The most usual form is inherited via a combination of genes, but a rare form is caused by a dominant allele. There is much variability as to how much babies are affected: in some, only the lip is cleft (hare lip); in others, the palate is cleft right to the back of the mouth. Surgery can repair the damage to a great extent, surgical techniques in this area having greatly improved. Boys are more commonly affected than girls; about 1–2 per 1000 white babies have cleft lip/palate, but the frequency is lower in blacks.

Cline, a gradient of variation in a continuously varying character, over a geographical range. It can also be a gradient in the frequency of a phenotype or gene.

Clonal selection theory, an idea put forward in 1951 by the Australian immunologist Sir Macfarlane Burnet to explain how the >immune system recognizes antigens and produces the appropriate antibodies. He proposed that, during foetal life, the body's lymphoid tissue produces numerous clones of cells, each specialized for making a particular antibody. The clones remain small and inactive until 'their' antigen arrives in the body, whereupon the clone becomes active and multiplies to produce large quantities of the antibody (*see* B lymphocyte). Having been stimulated once, the clone persists at a more numerous and active level than clones that have never been activated; this explains how immunity against a disease can be acquired and retained. Burnet also proposed that, during early foetal life, the cell lines that would have made antibody against the body's own tissues are selectively destroyed; if this is not correctly done, an auto-immune disease occurs.

Clone, a group of organisms that are genetically identical because they are derived from a single cell. Clones occur in nature, as the result of >vegetative propagation. Experimentally, clones have been made of toads: they are especially suitable because the nucleus of the single-cell embryo (one egg in the spawn) is easily removed and replaced by one from any cell in the body of the toad to be cloned. The embryo develops as it normally would, except that its entire genetic apparatus is identical to that of the toad that donated the nucleus. The number of identical toads that can be produced in this way is limited only by the patience of the experimenter. The same system could work in any animal – humans included – provided there is access to the single-celled fertilized egg and there is the technical capability to enucleate it.

A simpler method that has been used in sheep and mice is to take the four-cell stage of the embryo, break it up into four individual cells (each genetically identical, and still with the full potential to develop into a complete individual), keep them viable and allow them to complete their development (mammalian embryos have to be reimplanted, whereas those of toads and other creatures do not). This could also be done with humans, and is in fact similar to the process that happens in nature when the two-celled embryo splits in two and develops as identical twins, who are technically a clone. (Identical quadruplets are possible too, if the embryo splits up at the four-celled stage.)

The word 'clone' can also be applied to a group of cells which are all derived from a single parent-cell; this could be either in the body of an organism or in tissue culture (*see* monoclonal antibodies). *See also* gene cloning.

Coadaptation, the simultaneous selection of two or more genes which when present together give increased fitness. The resulting gene complexes become progressively established in the gene pool of the population or species. Coadaptation can also mean simultaneous evolution of two different species who interact ecologically to their mutual benefit.

Co-dominant, term to describe two (or more) alleles which express their effect whether or not the other is present. The genes for A and B blood groups are co-dominant. *See also* semi-dominant.

Codon, a sequence of three bases in DNA or messenger RNA, forming a 'word' in the >genetic code. *See also* protein synthesis.

Colchicine, a substance derived from the autumn crocus and used in genetic research. If it is added to a culture of cells, it blocks the process of >mitosis; the chromosomes all become arrested at metaphase, the stage at which they are thickened and easily visible.

Colour blindness, inability to distinguish between certain colours, a genetically determined condition. There are several types of colour blindness in humans. The most common are the two forms of red–green colour blindness, both of which are due to recessive mutant alleles on the X chromosome. About 8 per cent of men in European populations have one or other of these genes and are therefore unable to distinguish red from green, while about 6 out of every 1000 women have this kind of colour blindness. Other populations have lower frequencies; e.g. 2.3 per cent of Mexican and 1.7 per cent of Congolese men are colour blind. A much rarer form of colour blindness is the total inability to see colours, which is inherited as an autosomal recessive.

Competitive-exclusion principle, the idea that two species that inhabit exactly the same ecological niche cannot co-exist indefinitely. Though they may do so for a time eventually one will drive the other out by being a more successful competitor for the available resources. The few examples found in ecological genetics that appeared to contravene this principle have turned out to involve species whose ecological niches were not after all exactly the same.

Complement, a group of proteins involved in the >immune response. After antibody molecules become bound to the invading antigens, complement is produced and facilitates the dissolving of the antigenic cells. There are nine separate steps (each involving an enzyme) in what is known as the 'classical pathway of complement activation'. After complement has reacted with the antibody–antigen complex, it is said to be 'fixed'.

Complementary DNA (cDNA), a DNA molecule that has been artificially produced, not by replication, but by reverse transcription of a strand of RNA. *See* reverse transcriptase.

Complementary genes, two or more genes that are not alleles (i.e. they are at different loci, and may or may not be linked) that must work together to produce the effect in the phenotype. An example would be genes for two different enzymes, both of which are needed to synthesize a particular protein. Complementary genes can be either dominant or recessive.

Complementation test, a technique which tests whether two mutants are alleles of each other or are at separate loci. For example, let strain A be homozygous for the recessive mutation *a* (all individuals have genotype *aa*), and strain B for *b* (all individuals *bb*). If they are crossed and the genes are not alleles (i.e. they are at separate loci), the result will be a normal phenotype, because at each of the two loci, the mutant is opposite the >wild-type allele, i.e. the genotype is *a*+/+*b*. If *a* and *b* are alleles, the result will be a mutant phenotype, because the genotype will be *ab*++, i.e. at that one locus there is no wild-type allele, only *a* and *b* which are both mutants. The phenotype of these hybrids is not necessarily the same as that of either of the strains A and B, because although *a* and *b* are both recessive to the wild type one may be semi-dominant over the other.

Congenital defects, any of the various things that go wrong with a baby's development that are due not to genes but to the environment in the uterus during pregnancy. These include the results of infection (e.g. the deafness and blindness caused by >rubella), deficiency in the mother's diet, or contact with harmful chemicals. Such defects are not inherited if the affected child grows up to have children of its own (they are an example of >acquired characteristics). In some cases, there is an interaction between an environmental cause and a genetic predisposition to the defect; this seems to be true of >spina bifida, which is related to maternal diet but also has a genetic element. In Britain congenital defects account for some 18 per cent of all admissions to hospital for paediatric treatment.

Conidia, asexual spores produced by the haploid-phase >hyphae of some fungi. Conidia from type *A* hyphae can fertilize type *a* hyphae, and vice versa, but conidia cannot fertilize hyphae of their own mating type.

Conjugation, the form of sexual reproduction that single-celled organisms, including bacteria, sometimes use (as discovered by >Lederberg and >Tatum). Two cells form a bridge between their cytoplasm and exchange genetic material; this is carried on plasmids (extrachromosomal DNA rings) rather than on the chromosome. In >*Escherichia coli*, the transfer of DNA is one-way only, but in other genera, it is reciprocal. Conjugation can also occur between different genera (e.g. *E. coli* and *Salmonella*).

Consanguinity, relatedness. The amount of genetic relatedness between individuals is measured by the coefficient of >relationship. The word is from Latin, meaning 'shared blood': *see* p. 9 for the persistence of this idea. *See also* genetic distance.

Consensus sequence, a sequence of DNA that is repeated in a number of different places in the genome of an organism or in many different organisms, presumably because it serves a similar function each time. An example is the >TATA box.

Conserved sequence, a sequence of DNA that has varied little during evolutionary time. One example is the H3 and H4 types of histone. *See* molecular evolution.

Constant region. *See* C region.

Constitutive, a term used to describe an >operon or a gene that is more or less constantly transcribed, as opposed to an operon or gene that is under some form of >gene regulation. Also used for a constitutive enzyme, one that is constantly produced, as opposed to an inducible or repressible enzyme.

Containment, a term that embraces all the techniques for making sure that potentially dangerous viruses, bacteria, etc. are safely handled within the laboratory. Isolated areas, controlled air pressure, radioactive decontamination, and so on are used.

Continuous variation, variation between one individual and another such that the various phenotypes do not fall into distinct classes but are a continuum and can only be distinguished by some sort of measurement.

Much variation in biology is of this type, and it was originally believed that Mendelian inheritance could not explain it. In 1908 >Nilsson-Ehle carried out classic experiments with a cross of wheat that had seeds continuously varying from dark red to white, and showed that only two gene pairs were involved. Later Sir Ronald >Fisher proved mathematically that data on continuous variation can only be explained as being due to separate Mendelian genes interacting with each other and with the environment. Current thinking is that even complex continuously varying traits such as body weight are probably controlled by a small number of genes. *See also* polygenes.

Contraception, prevention of unwanted pregnancy. It is not an issue in genetics, unless it can be shown – as early proponents of eugenics believed – that contraception is only practised by persons of superior genotypes, leaving the inferior types to swamp the human race with their genes. If contraception is practised by large numbers of people irrespective of their genotype (as of course is the case) it is of no significance genetically.

Convergence, evolutionary change that brings two unrelated species to resemble each other, either in their overall >phenotype or else in one particular trait. An example is the black swan of Australia, which is remarkably similar to the white swans of the Northern Hemisphere, which have a completely different ancestry. *See also* parallelism.

Conversion, the phenomenon in which one allele is converted into another. It can happen during either mitosis or meiosis but is more easily detected in the latter. If the two alleles are *A* and *a*, the products of meiosis should be two *A* gametes and two *a*s. If gene conversion occurs, the product is three *A*s to one *a* (or the other way round). It may be that the DNA polymerase molecule making the new copy switches temporarily from one homologous chromosome to the other.

Copy error, a >mutation that results from an error of copying during the replication of one strand of DNA from another.

Correlation, coefficient of, a measurement of how closely associated two paired sets of variables are. In genetical applications, the variables might be the same trait measured in pairs of related individuals or two traits measured in a sample of individuals. It is defined as the >covariance of the two variables divided by the product of their own >variances. The values range from +1 (total positive correlation: when one variable goes up, the other always goes up too) to 0 (no correlation: variation in one is random with respect to the other) and then to −1 (totally negative correlation: when one goes up, the other goes down), with all points in between being possible. These correspond to what one would see if the two variables were plotted against each other on a graph: +1 would be seen as points lying around a line rising diagonally up from lower left to upper right; 0 would be a scatter of points at random; −1 would be points around a line falling diagonally from upper left to lower right; and other values would be lines with slopes other than the perfect diagonal. The method of calculating correlations was devised by >Galton.

Correns, Karl Erich (1864–1933), German botanist, one of the three rediscoverers of Mendel's work. In 1899 Correns was working at the University of Tübingen on plant hybridization when his experiments with peas produced the same ratios in the hybrids as Mendel's had done, and Correns made the same theoretical deductions. A few weeks later, he found Mendel's original paper, and when de >Vries's paper proving the same theory came out in 1900 Correns was ready with his own. Of all Mendel's rediscoverers, Correns had the best grasp of the theory of heredity. He was the first to prove that segregation of the alleles does not depend on one being dominant and the other recessive, and in 1903 he predicted that sex would be found to be inherited as a Mendelian factor – as of course it is.

Coupling. *See cis* configuration.

Covariance, a statistical technique to measure the association between two variables (continuously varying). These might be two traits in the same individual (e.g. intelligence and shoe size) or the same trait in related (or otherwise paired) individuals. The covariance is the product of the deviations from the two means, averaged over the number of observations. With no association, the individual values will vary from the mean in opposite directions or not at all as often as they vary together, giving a covariance value of about 0 (e.g. if very intelligent people have large feet as often as they have small); but if there is association, the covariance value increases. *See also* correlation, coefficient of.

Creationism, the belief that the Earth and the life forms on it were made by a creator. This could have happened all at once, as in the biblical account, or on more than one occasion, as Cuvier thought (*see* p. 14). A modernized version is that the first living cell was created, and that all life has been evolved since then by the creator using >natural selection as its instrument. As recently as 1981 the state of Arkansas passed a law requiring 'balanced treatment for creation-science and evolution-science', meaning that fundamentalist creationism had to be taught in schools alongside Darwinism; the law was overturned the following year after appeals by teachers.

Crick, Francis Harry Compton (1916–), British molecular biologist, joint discoverer of the structure of DNA. Crick trained as a physicist and served in World War II in naval weapons research, but after the war he became interested in biology and joined the Medical Research Council's Molecular

Biology Research Unit then located in the Cavendish Laboratory at Cambridge. His research programme was in X-ray diffraction studies of protein structure, but when the geneticist James >Watson arrived, with his enthusiasm for finding out what genes were really made of, Crick turned his imaginative scientific mind to the problem. They used what clues were available – >Avery's identification of DNA as the genetic material; >Chargaff's rules on the ratios of the bases in DNA; >Franklin's X-ray diffraction photographs – but their final success in finding the double helix came when they built a scale model out of pieces of cardboard and wire. Crick and Watson's paper in the scientific journal *Nature* in 1953 described their finding is one of the classics of scientific literature, and is only one page long.

Crick's brilliant originality led him to further fundamental insights in to molecular biology, most notably the >central dogma, which states the following progression:

DNA → [transcription] → RNA → [translation] → protein

With Sydney >Brenner, Crick also did much of the work in establishing that the >genetic code is in triplets (*see also* wobble). He continued to make contributions to the theory of molecular biology, including proposing the idea of >selfish DNA. Crick shared the 1962 Nobel prize for physiology or medicine with Watson and Maurice Wilkins. He continued to work at the MRC laboratory until 1977, when he went to the Salk Institute at La Jolla, California. *see also* pp. 28–32.

- J. D. Watson, *The Double Helix* (1968): a good read, although Watson's portrait of Crick is ungenerous.

Cro-Magnon Man. *see* human evolution.

Cross-over value, the number of individuals who show recombinant types (as the result of >crossing-over) expressed as the total number of individual offspring in a cross. *See also* map unit.

Crossing-over, the reciprocal exchange of segments between non-sister chromatids during >meiosis. This happens during the pachytene stage; the homologous pairs of chromosomes lie alongside each other, and each chromosome is doubled longwise into two identical chromatids. One chromatid of one chromosome may at any point at random join up with the corresponding point on a chromatid of the homologous chromosome, and exchange all alleles from that point to the end of the chromosome (or to the next point of crossing-over, because it can happen more than once to each chromosome pair at each meiosis). This process results in >recombination of alleles between the maternal and paternal chromosomes. For instance, if the alleles on one chromosome were H L and on the other were h l, and there was a cross-over between these two loci, the result would be one chromatid with $H\ l$ and the other with $h\ L$ (see diagram). Note that there are also the two chromatids which were not involved in the cross-over; these retain the original arrangements, $H\ L$ and $h\ l$. The point at which the chromatids cross is called a chiasma.

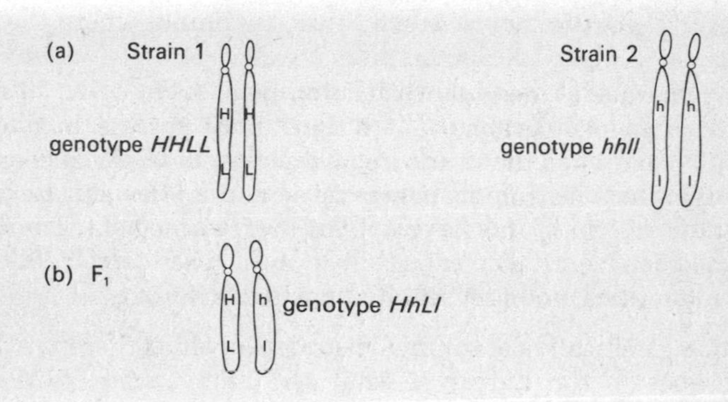

(c) During meiosis the chromosomes double into twin chromatids, and crossing-over and exchange of genetic material takes place:

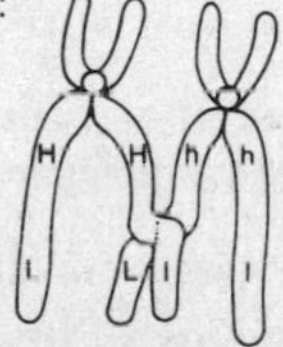

(d) After the final meiotic division, there are four chromatids: *HL*, *Hl*, and *hl*

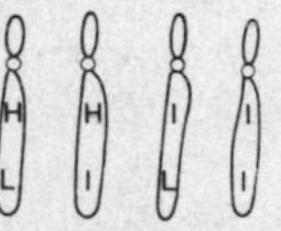

(e) The chromatids go into gametes, and the ones with genotypes not found in the parental strains (2nd and 3rd from left above) are called recombinants.

The further apart on a chromosome two genes are, the greater is the probability that a cross-over will occur between them, and this fact is used to make >gene maps. *See also* intrachromosomal cross-over.

Cultivar, a plant variety that does not occur in nature. A cultivar can be produced either by selection within a wild-derived strain, or by crossing either within a species or between species or genera.

Cultural evolution. *See* exogenetic inheritance.

Cystic fibrosis, a human disease inherited as a recessive. The clinical effects are lung damage (always) and swelling and malfunction of the spleen, liver or pancreas. If untreated, 75 per cent of babies die within their first year, of lung complications. If they are given treatment to prevent lung damage, sufferers live on average until their early 20s. Cystic fibrosis is the commonest genetic disease in Europeans, occurring in about 1 in 2000 births; this implies that about one person in 23 is a heterozygote. The frequency is much lower in China, where the disease occurs at a rate of 1 per 100,000 births.

The gene responsible was known by the early 1980s to be situated on chromosome 7, and was found, isolated and cloned in 1989. It differs from the equivalent normal gene in having a deletion of three base-pairs (i.e. one >codon) so that one >amino acid is missing from the gene product; the biological implications of this have not yet been worked out. The fact that the gene has been identified means that in most cases carriers of the gene can now be identified, and prenatal diagnosis carried out.

Cystinuria, a genetically determined disorder in which there is a defect in the membranes in the kidney so that the amino acid cystine is not reabsorbed from the urine as normal. Stones made of cystine form within the kidneys and can ultimately destroy them. The reabsorption defect is due to a recessive gene, and about 1 in 12,000 babies is born with the disorder. Cystinuria was one of the first four Mendelian traits in humans that Sir Archibald >Garrod discovered in 1903.

Cytogenetics, the study of the number, behaviour and morphology of chromosomes as a means of understanding the genetics of an organism.

Cytoplasmic inheritance, genetic inheritance via the DNA-carrying >organelles in the cytoplasm, i.e. the chloroplasts (plants only) and mitochondria. In mammals and many other groups, the sperm contributes virtually no cytoplasm, whereas the ovum is a large cell, so that cytoplasmic inheritance is exclusively maternal. It is not, however, an important feature in mammalian genetics. *See also* maternal inheritance.

Cytosine (abbreviation: C), one of the four nitrogenous bases which form the core of DNA and RNA. It pairs with guanine (G). *See also* replication.

D

darwin, a unit of evolutionary change. A darwin is equal to a change (increase or decrease) in any one character by a factor of 2.7 per million years. As this is a very fast rate of change, most evolutionary change is expressed in millidarwins.

Darwin, Charles Robert, (1809–82), British naturalist, co-discoverer of the principle of evolution by natural selection. Born into a distinguished family (his father's side were outstanding scientists, including his grandfather Erasmus Darwin, and his mother was one of the Wedgwoods of pottery fame), Darwin was himself an unpromising pupil who studied medicine at Edinburgh and theology at Cambridge but graduated without honours. He had spent his whole time at university following his passion for the study of nature, and he was a formidably accomplished collector. When he heard that there was a place for an unpaid naturalist on board a naval exploration ship, he was determined to go, in spite of his family's opposition. Darwin boarded HMS *Beagle* in December 1831; the voyage lasted until October 1836, and included visits to South America, Australia, Tasmania, New Zealand, Tahiti, Mauritius, and many other places. Darwin was as interested in geology as in anything else, and in his *Structure and Distribution of Coral Reefs* (1842) he was the first to put forward a theory (still held) of reef formation. For an account of his arrival at the theory of evolution by natural selection, see pp. 15–18.

After the momentous publication of *On the Origin of Species* in 1859, Darwin wrote many books extending his basic theory, though his biological thought was always hampered by the fact that the mechanism of heredity was unknown (although it had, in fact, already been discovered by Mendel). *The Variation of Animals and Plants under Domestication*

(1868), *The Descent of Man* (1871), *The Expression of the Emotions in Man and Other Animals* (1872) and *The Effects of Cross and Self Fertilization in the Vegetable Kingdom* (1876) were the most important of his later works. His collected *Letters* to his many scientific friends contain many of his most interesting thoughts, including his remarkable speculation on the origin of life.

Darwin never held a paid post, being one of the last of the hitherto important 'gentleman-scientists'. His beliefs are summed up by his statement (1870) that 'I cannot look at the universe as the result of blind chance, yet I can see no evidence of beneficent design, or indeed of design of any kind, in the details.' *See also* Wallace; Huxley; Lyell.

Deafness, the inability to hear because of *(a)* damage or malformation in the structure of the ear or *(b)* a neurological defect. Deafness has many different causes, but some forms are inherited. In one, the baby is born deaf; this is caused by a recessive gene and occurs in about 1 per 10,000 births. In another, inherited as a dominant, deafness sets in at about the age of one year; this is about twice as common. More frequent still is the form of deafness that sets in during early adult life, which is caused by a dominant allele and affects about one person per 1000. Persons deaf from birth or infancy are also mute.

The genetics of inherited deafness has been well studied because of the frequency with which deaf adults intermarry. This has led to the conclusion that there at least two different recessive genes that cause total deafness because sometimes two parents, both with recessively inherited deafness, have children who are normal, which could not happen if the parents were both homozygotes for the same gene.

Degenerate code. *See* redundant code.

Degrees of freedom, the number of independent variables that can vary within a statistic or total. If one has a sample of individuals to assign to four classes, there are three degrees of freedom: there might be any number in the first class (first degree used up), and any number in the second (second degree used up), but once the number in the third class has been assigned (third degree used up), there is no freedom for the last class to vary – it must simply be the remainder (even if zero). The general formula is $n-1$.

Deletion, the loss of a part of a chromosome. The lost portion may be as small as a single base-pair or as large as a whole arm of the chromosome. Deletions can occur at any point along the chromosome, though those involving loss of an end are most common. A deletion is one form of >mutation, and is likely to be lethal when homozygous (i.e. it occurs on both chromosomes) but viable when heterozygous. The latter is true because the genes on the normal chromosome can do the work of the ones deleted on the opposite number; recessive genes will be expressed, and are said to be 'uncovered' in such a situation. Several syndromes in humans are caused by deletions; *see* chromosome abnormalities. *See also* deletion mapping.

Deletion mapping, a technique for mapping the position of genes on a chromosome by using variants with >deletions on that chromosome. The technique depends on the deletions being visible under the microscope, and has therefore been used most extensively with *Drosophila* (because of its large chromosomes). A recessive mutation can be located because when a strain carrying this mutant is crossed with a strain with a deletion, the recessive will be expressed if it is uncovered, i.e. if it is on the stretch of chromosome missing by deletion. The accuracy of the mapping increases if there are strains available with different overlapping deletions or if deletions of short sequences have been identified. *See also* genetic map.

Deme, a local interbreeeding unit within which there is unrestricted random mating; part of a population.

Demography, the study of the dynamics of populations. Among the relevant variables are birth and death rates, age structure, immigration and emigration rates, some of which are genetically influenced and/or have an impact on the genetic composition of the population (which can be human or any other species).

Denaturation, generally the destruction of a macromolecule's three-dimensional structure. With reference to DNA, it means the breaking down of the double-stranded form into single strands or fragments, usually by heating. The process is reversible, the two strands reassociating themselves as they cool (*see* annealing). Proteins can also be denatured, but they can very seldom regain their original structure.

Dendrogram (also known as a 'phylogenetic tree'), a diagram in the shape of a tree which shows the ancestry of different species (or other taxa). The trunk of the tree is the original ancestor, and each time a new species is formed, it splits off as a branch. The lower down a branch occurs, the earlier the species. *See also* family tree.

Density-dependent factors, a number of mechanisms by which a population's density in relation to its environment is self-regulating. When density is too great, organisms may become less fertile or even cease to breed, mortality may increase or there may be emigration, until the population declines to a level where its density is right. Conversely, at low density fertility tends to increase and mortality to decrease.

Deoxyribonuclease (DNase), one of a group of enzymes that can cut a DNA molecule into fragments by breaking the sugar–phosphate bonds. It may work by cutting off one base at a time from one end (when it would be an >exonuclease) or by cutting at a specific point along the sequence (>endonuclease).

Deoxyribose, the sugar that forms part of the structure of >DNA.

Deoxyribonucleic acid. *See* DNA.

Dependent character, a character (i.e. trait) that occurs only if another character is also present. For example, there could be a trait for 'curly horn' in sheep that is expressed only if the individual sheep also has the gene for horns. Sheep having the hornless gene are unable to show the 'curly' type whether or not they carry the gene for it. *See also* sex-limited trait.

Derepression, activation of a gene when the >repressor protein is removed. *See* gene regulation.

Developmental genetics, the study of the genetic control of embryogenesis, growth, regeneration and degeneration. A key concept is the idea that a developmental gene is not a 'description' of, say, a tall horse; rather, it is an 'instruction' for how to make a tall horse. Many developmental genes appear to work by altering the timing or duration of production of some critical substance.

Diabetes mellitus, a metabolic disorder in which sugar, i.e. glucose, is not properly metabolized owing to the lack of the hormone >insulin. Early-onset (or insulin-dependent) diabetes is an >auto-immune disease in which the insulin-producing cells in the pancreas are destroyed; its pattern of inheritance is not clear, but it is closely associated with HLA types DR3 and DR4. The maturity-onset form also has a genetic component but cannot be attributed to a single gene. The gene for insulin in humans has been found (on chromosome 11), but it does not seem that a mutation in this gene is the cause of diabetes. Diabetes affects up to 1 per cent of adults in white populations.

Diakinesis, one of the stages in >meiosis (one of the two processes of cell division). *See* diagram under meiosis.

Dicentric, a chromosome with two >centromeres. A dicentric is an abnormality that tends to break up because its two ends try to go in opposite directions during cell division. *See* acentric.

Dihybrid, an organism that is heterozygous at two different loci. The word is also used for a cross between two such individuals. In such a cross with independently assorting characters (with dominance of one allele over the other at each locus), there are four phenotypes among the resulting offspring, in the ratio 9:3:3:1, as shown in the diagram (known as a >Punnett square).

Assume that the cross is between two rabbits: *A* is the allele for black fur, and is dominant to *a* for white; *Z* is the allele for long ears and is

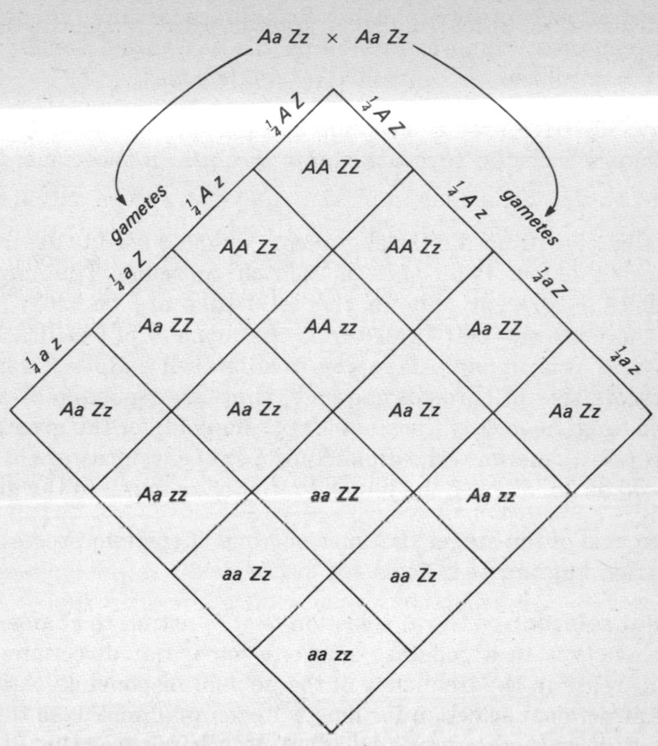

$\frac{1}{16}$ *AA ZZ* + $\frac{2}{16}$ *AA Zz* + $\frac{2}{16}$ *Aa ZZ* + $\frac{4}{16}$ *Aa Zz* = $\frac{9}{16}$ *black, long ears*

$\frac{1}{16}$ *aa ZZ* + $\frac{2}{16}$ *aa Zz* = $\frac{3}{16}$ *white, long ears*

$\frac{1}{16}$ *AA zz* + $\frac{2}{16}$ *Aa zz* = $\frac{3}{16}$ *black, short ears*

$\frac{1}{16}$ *aa zz* = $\frac{1}{16}$ *white, short ears*

dominant to *z* for short ears. Both parents (with genotypes *Aa Zz*) are black with long ears.

Every possible genotype combining the four alleles occurs, and there are 16 of them. The most frequent (at 4 out of 16) is the dihybrid – double heterozygote (*Aa Zz*) – like the two parents, and the least frequent (1 out of 16) is double homozygote (*aa zz*). This is exactly comparable to the position in the basic or >monohybrid cross, only with the numbers squared. *See also* hybrid; trihybrid.

Diminution, the reduction in the number of chromosomes in somatic cells. This is a feature of some nematode worms and some species of flies.

Dimorphism, the existence in a population of two distinct types, genetically determined. One example is sheepdogs, which are either rough-coated (like Lassie) or smooth. A much more common situation is the existence of more than two types, which is called >polymorphism. The most important form of dimorphism is sex (*see* sexual dimorphism), though it is a special case, being controlled by the sex chromosomes; most instances of dimorphism are controlled by single genes.

Dioecious, adjective used to describe a plant species that has male and female flowers borne on separate plants. *See also* monoecious; flowering plants.

Diploid, adjective (from the Greek *diploos*, 'double') used to describe a cell or organism that has two copies of each chromosome. The single set of chromosomes is present only in the cells that are gametes (ovum or sperm), whereas every cell formed from the moment of fertilization has a double set, i.e. it is diploid. All species of animals are diploid; some plants have multiple sets of chromosomes (i.e. they are >polyploid) and some yeasts and fungi have only the single set (>haploid) for the greater part of their life cycle. Sometimes the diploid number of chromosomes of a species is expressed as $2n$, n being the haploid number as found in the gametes.

Diplotene, one of the stages in >meiosis (one of the two processes of cell division). *See* diagram at meiosis.

Directional selection, natural selection that is acting to change the prevailing phenotype in a population in one particular direction. There is then an increase in the frequency of the gene(s) responsible. An example would be directional selection for longer fur on mammoths as the climate grew colder. *See also* disruptive selection; stabilizing selection.

Discontinuous variation, variation between one individual and another such that each type falls into a separate class. This was the type of variation studied by Mendel, who looked at such discontinuous traits as flower or seed colour in peas. *See also* continuous variation.

Disjunction, the separation of homologous chromosomes during either of the two processes of cell division, >meiosis or >mitosis. *See also* nondisjunction.

Disposable soma theory, a theory that attempts to explain ageing. Since, by definition, natural selection favours reproduction rather than an individual's survival *per se*, therefore an organism cannot afford (in bioenergetic terms) to spend more effort in repairing its body than it does in reproduction. This results in the selection of organisms that cease to carry out repairs (i.e. show senescence, or ageing) once their reproductive capacity has diminished. *Soma* is the Greek for 'body'; the name of the theory implies that the organism's body is expended in favour of reproductive performance.

Disruptive selection, >natural selection that is acting to maintain two or more phenotypically different types in a population. The sets of genes responsible for the selected phenotypes become more common, with any intermediate genetic combinations selected against. An example of disruptive selection is >mimicry: in a (palatable) species of butterfly the normal phenotype is favoured, and so are phenotypes that are exact mimics of the other (unpalatable) species, but phenotypes that are only halfway to mimicking the other species are useless and are selected against. *See also* directional selection; polymorphism; stabilizing selection.

Distribution, principle of, the idea that related species that are geographically near to one another are more alike than those that are far apart. This not-surprising observation was one of the factors that led Wallace and Darwin to their theory of evolution. The creationist explanation was that God had placed each creature in the habitat that suited it best. The principle of distribution is the geographical equivalent of the geological principle of >succession.

Dizygotic twins, non-identical ('fraternal') twins. As the word implies, these twins develop from two separate >zygotes: there were two ova and two sperm and the twins are therefore genetically different individuals, no more alike than any other siblings. They can even be of different sexes. It is curious that dizygotic twins tend to 'mirror' each other in things like which side the hair parts on or which is the preferred hand: a right-handed and left-handed pair of twins is quite usual. Like-sex dizygotic twins have been used in >twin studies as a basis for comparison with identical twins (*see* monozygotic twins). *See also* twinning rates.

DNA (deoxyribonucleic acid), the substance of which genes are made. It has three crucial properties: *(1)* it stores and faithfully transmits information; *(2)* it copies itself accurately; and *(3)* it can occasionally mutate, and when it does, the >mutation is inherited. DNA was identified as the genetic substance by Oswald >Avery in 1944, and though his discovery was at first ignored (*see* p. 28), it was confirmed by Alfred >Hershey in 1952.

The three-dimensional structure of DNA, defined the following year by Francis >Crick and James >Watson (*see* p. 28), is a double helix. It has been likened to a spiral staircase: the 'banisters' are composed of alternate phosphate and sugar molecules (the sugar being deoxyribose), but the key to DNA is in the 'steps' made of pairs of nitrogenous bases locked together crosswise. Base A (adenosine) on one side must be joined to T (thymine) on the other or vice versa, and C (cytosine) must be joined to G (guanine) or vice versa. This means that DNA is able to self-replicate: the double-stranded molecule splits lengthwise, and each single-stranded half is able to reconstitute its opposite strand by following the base-pairing rules. This is what happens every time a cell divides, when the complete set of chromosomes has to be copied.

The information-carrying capacity of DNA also comes from the bases; they are 'read' as if they were letters making up words, each three bases long. These words give the information needed for building proteins and

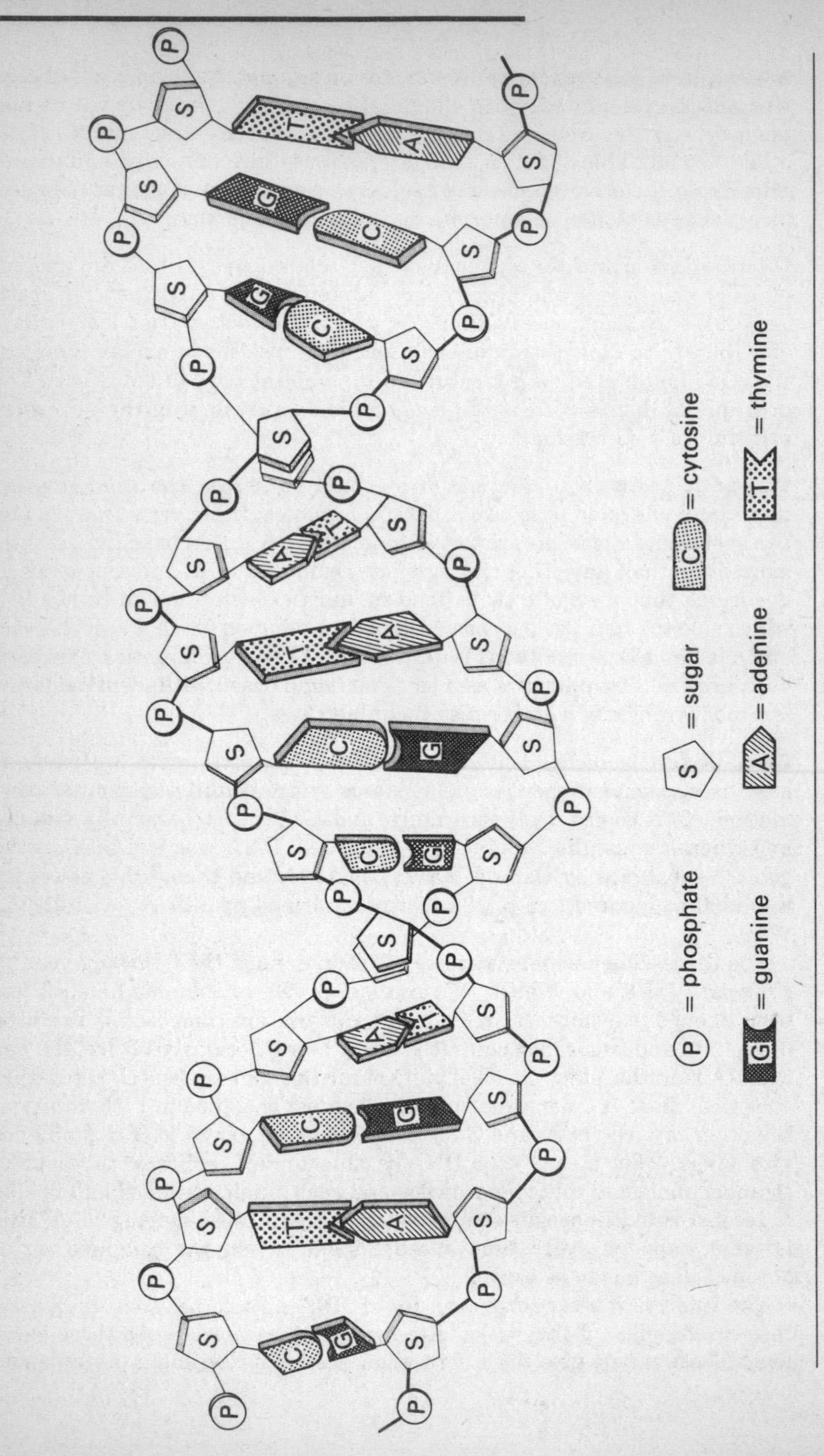
P = phosphate
S = sugar
C = cytosine
G = guanine
A = adenine
T = thymine

for organizing the activity of the cell. *See also* central dogma; chromosome; genetic code; protein synthesis; replication.

DNA amplification. *See* polymerase chain reaction.

DNA cloning, the technique of isolating, maintaining and faithfully reproducing many identical copies (i.e. clones) of a piece of DNA such as a gene. This can be done by the transfer of DNA from an external source into a host bacterial cell via a vector. The extraneous DNA may be prepared by (*a*) physical breakage of complete chromosomes or large DNA molecules by stirring at 0°C (32°F); (*b*) digestion by a restriction endonuclease; (*c*) transcription from an RNA template by reverse transcriptese (cDNA cloning); (*d*) direct synthesis. The artificially produced DNA resulting from this is called >recombinant DNA.

DNA library, a collection of fragments of cloned DNA (*see* DNA cloning) that together represent all of the >genome of the organism that the DNA came from. A cDNA library can be constructed by transcribing the mRNA to DNA using reverse transcriptase, and then cloning. This provides a library of the genes that are active in the cells from which the mRNA was isolated.

DNA repair, any of the processes by which the correct sequence is restored after a mutational change in DNA. One is *photoreactivation*, which repairs the typical erroneous T–T base-pairs induced by exposure to ultraviolet radiation, and is carried out by an enzyme that utilizes visible light. So-called *dark repair* is similar but carried out by enzymes that work best in the absence of light. In humans the DNA repair system probably consists of the products of seven separate genes. For example, the disease xeroderma pigmentosum results from a failure to repair DNA damage in skin cells, due to homozygosity for a recessive mutant allele at any one of these seven loci. If DNA repair were 100 per cent efficient, there would be no transmission of mutations, and therefore no evolutionary progress. *See also* mismatch repair; proofreading; repair synthesis.

Dobzhansky, Theodosius (1900–75), Russian-born American geneticist; professor of genetics at Columbia University, New York for many years. Dobzhansky researched the genetics of *Drosophila* in wild populations, and his findings provided the best illustrations of the Darwinian hypothesis. He was an excellent writer, responsible both for standard textbooks on genetics and evolution and for such popular works as *Mankind Evolving* (1962). Dobzhansky was highly sceptical of eugenics, and especially of the idea of >sperm banks.

Domestication, the transformation of wild plants or animals into forms that are useful to humans; the first applied genetics. Dogs may have been domesticated in China by *Homo erectus* half a million years ago. The first plants to be domesticated were primitive wheats (emmer and einkorn), found on sites dated to about 13,000 BC in Israel, with true wheat first found in Syria and dating from about 9000 BC. The domestication of sheep and goats dates from about 11,000 BC and took place in the Middle East; cattle were domesticated rather later, and in North America, the wild turkey was domesticated from about 5000 years ago. Yeast and the silk worm have both been domesticated since ancient times.

Domestication always involves selective breeding, with consequent reduction of the gene pool of the animal or plant involved. None the less, there are impressive numbers of different varieties of each domesticated organism: consider the domestic dog, all varieties of which are members of the same species, *Canis familiaris*. The sort of change that occurs under domestication (e.g. increased size) was one piece of evidence that suggested the theory of evolution to Lamarck, Wallace and Darwin.

Dominant, an allele that expresses its effect even when it is present only in a single dose. A familiar example is brown versus blue eyes: if the allele for brown eyes is *B* and that for blue eyes is *b*, then a person who has the genotype *Bb* will have brown eyes, because *B* is dominant to *b*. The opposite (*b* in this example) is called >recessive; alleles can also be >co-dominant, >semi-dominant or >overdominant. The notation using capital letters for dominant alleles was devised by Mendel, who was the first to explain the dominance effect (though it had been observed for literally ages).

Dosage compensation, a mechanism to equalize the genetic effects of having two X chromosomes in females compared with one in males. In mammals this is accounted for by the total inactivation of one X chromosome in each cell (*see* Lyon hypothesis). In *Drosophila* compensation does occur but both the Xs retain some of their activity; the mechanism is not understood.

Down's syndrome, a condition in which a baby is born with a number of defects including mental retardation and heart abnormalities; the typical flattened face gave rise to the syndrome's former name 'mongolism'. Down's syndrome children are notably sociable and affectionate, and the less severely affected can go to ordinary schools and later take sheltered work. About 50 per cent of those with Down's syndrome die before the age of five, and the rest mostly in their teens or twenties, from heart or lung complications. Boys with Down's syndrome do not grow up to be fertile, but girls do, and their babies are either normal or have Down's syndrome. The disorder is due to the presence of three copies (*see* trisomy) of chromosome 21 (hence the alternate name for the syndrome: trisomy 21). There is a strong association with maternal age, the risk of a baby being born with Down's syndrome increasing from about 1 in 2200 when the mother is 20 to about 1 in 40 when she is 45; none the less maternal age cannot account for all the cases of Down's syndrome, and is not a factor in about 40 per cent of them. It has been estimated that of all the foetuses conceived that have trisomy 21, twice as many are lost through spontaneous abortion (miscarriage) as are carried to term. The condition can be diagnosed prenatally, and it is now routine to offer prenatal diagnosis to pregnant mothers aged 35 or more.

Downstream, further ahead on a DNA molecule, in respect of the direction in which the sequence is being read. *See* replication.

Drosophila, a genus of fruit fly, used extensively in genetic research. There are 1467 species, found all over the world in warm countries. The species used in laboratories is *D. melanogaster*, and the one whose population genetics has been most studied (*see* Dobzhansky, Theodosius) is *D. pseudoobscura*. The typical *Drosophila* is 4–5 mm long, dark brown, with two wings. In the wild, its diet is usually rotting fruit. Generation time (i.e. from fertilization of egg to mature adult) is 2 weeks, and the normal lifespan is 4–5 weeks. In the laboratory flies are kept in bottles, which can contain up to 4000 individuals, and are fed on a solution of yeast, sugar and other nutrients. For experiments on gene frequencies, whole 'populations' of flies are needed, and these are kept in population cages, 30 cm cubes made of transparent plastic and containing up to 10,000 flies. There are four reasons why *Drosophila* became one of the main organisms for genetic research: (*1*) it is easy to keep; (*2*) its generation time is short; (*3*) it has only four pairs of chromosomes; and (*4*) it has 'giant' chromosomes in its salivary gland cells which are easily studied microscopically. There is an immense literature on the genetics of *Drosophila*, in which over 3000 mutants have been identified and mapped.

Duchenne muscular dystrophy, a human disease in which there is progressive weakening of the muscles, leading to death at about the age of 20. The cause is an X-linked recessive gene, so only boys are affected, although some of the female carriers do show some signs of progressive muscular weakening. The disease occurs at rate of about 2 per 10,000 male babies, and in about two-thirds of these cases the gene has been inherited from a carrier mother, while in the rest it has arisen as a new mutation. A woman who knows she is a carrier, because she has already had an affected son, can have prenatal diagnosis to ascertain the sex of her subsequent babies; she may then choose to abort any males, but half of these would have been normal.

Boys with Duchenne muscular dystrophy are never well enough to be fertile, so the gene dies out each time it is manifested in a boy, although it can travel indefinitely down generations in females. The gene itself was located by reverse genetics, and has been found to be exceptionally large: at 2 million base-pairs, it accounts for over 1 per cent of the X chromosome and is one of the longest genes in the human genome. The protein encoded by this gene is called dystrophin, and it is believed to be involved in muscle contraction; boys with the mutant gene do not have this protein.

Duffy blood group system, a system of genetically determined antigens on human red blood cells (discovered in 1950). There are three alleles: *Fy*, Fy^a and Fy^b. *Fy* is at very high frequency in black populations in Africa (and in people of African descent everywhere); some African populations have close to 100 per cent frequency of this allele. The reason for this very high incidence seems to be that the allele gives protection against one form of malaria (but without the deleterious effects that come with sickle-cell anaemia, also associated with protection against malaria).

Duplication, a type of chromosomal mutation in which a segment of a chromosome is repeated. If the duplicated segment is next to the original one, it is a tandem duplication, but the duplication may end up anywhere, even on a different chromosome or not attached to a chromosome at all.

Dwarfism, the condition of having very short stature, usually due to excessively short limb bones. The most common form is >achondroplasia. Some other forms are due to non-production of growth hormone, and children born with these can be treated with this hormone. Formerly the growth hormone had to be obtained from the pituitary glands of human corpses, but now the gene has been cloned and so human growth hormone can be produced in bulk by bacteria.

E

Early gene, a gene in the genome of a bacteriophage or an animal virus that is transcribed early, during the time when the virus is reproducing itself within a bacterial cell. The products of early genes are necessary for regulating or transcribing the later genes necessary for the virus's reproductive cycle.

Ecad, an organism whose phenotype has been adapted to suit a particular environment. An ecad's adaptation is environmentally produced and therefore not heritable (though the idea that it was heritable was the mainstay of the theory of the inheritance of acquired characteristics).

Ecological niche, an available 'space' for a species where it does not compete with others for such resources as food and shelter. There can be many ecological niches within the same geographical location, the species finding their own spaces within it by having different diets or nesting requirements, or perhaps by occupying it at different times (day vs night, winter vs summer). *See* competitive-exclusion principle.

Ecology, the study of the interrelationships of species and their environment, including both inter- and intra-species relationships. Genetics is an important factor in ecological studies, as every ecological situation is fluid and represents mutual or competitive genetical adjustments being made by the species involved. This is why a geneticist would say that the study of ecology is really the study of evolution.

Ectogenesis, developing an embryo outside the body of the female. This is what happens in frogs, newts and others that have spawn developing into tadpoles, etc. If artificial ectogenesis were ever to be achieved in humans, the result would have more claim to be called a 'test-tube baby' than the babies produced by *in vitro* fertilization who are now so called. Partial ectogenesis for humans already exists at both ends of the process, in that eggs can be fertilized outside the body and the embryo grown for several weeks, and that premature babies as small as 470 grams (just over 1 lb) can survive.

Effective population size, the number of individuals in a population who contribute offspring to the next generation. It is considerably less than the absolute population size, in that some individuals are not fertile, and in some populations dominance hierarchies restrict the chance to breed to a few individuals. Very few real-life effective population sizes in animal species have been measured: some examples are about 10 for wild groups of the house mouse *Mus musculus* (with male dominance), and from 19 to 90 in a species of lizard (no male dominance).

Electrophoresis, a technique for identifying the different molecules present in a mixture, by placing them in a gel in an electric field; molecules such as proteins and DNA are negatively charged and therefore move towards the positive pole. Smaller molecules move through the gel faster than large ones, so that after a certain time they spread out and can be identified by how far they have moved, which can be seen when the gel is treated with an appropriate stain.

Electrophoresis can also be combined with other techniques. If the sample contains a radioactively labelled molecule, its position on the gel after electrophoresis can be found by autoradiography or scintillation counting. In immunoelectrophoresis, the samples are first separated and then identified by their reaction to specific antibodies.

Embryo, the organism in its early stages after fertilization. *See* embryogenesis.

Embryo transfer, the technique of artificially implanting early embryos into a surrogate mother. In humans, it is the key technique in >surrogacy. Embryo transfer is often used in cattle-breeding: a cow is given a fertility drug to make her go into multiple ovulation (i.e. produce more than one egg at the same time); she is artificially inseminated; and the resulting embryos are removed (and stored frozen if necessary) before being implanted, preferably two at a time, into surrogate mothers. In this way, the best cow in the herd can have seven or eight calves in a season instead of just one.

Embryo transfer is also being introduced for attempts to get rare or endangered species to raise more offspring, by getting a female of the rare species to go into superovulation and implanting the resulting embryos into females of a related but commoner species with a similar gestation time. There are practical problems. First, the stress caused to all the females may reduce their reproductive ability. Second, an infant brought up by a mother of the wrong species may not be able to breed with its own species.

Embryogenesis, the process of development of the organism from fertilization to birth. The details vary from one type of animal to the next; the following is the outline of the stages that the human embryo goes through.

After >fertilization, the egg immediately divides into two cells, four cells and so on. After about four days, it is called the morula and consists of a ball of 16 or more undifferentiated cells called blastomeres. At about five days the cells making up the ball differentiate and become the blastocyst, which comprises an outside layer (the *trophectoderm*) around a hollow centre containing the inner cell mass (totalling 36 cells at the beginning of this stage). After implantation into the wall of the uterus (which begins at about 7 days and is completed about 14 days after fertilization), the embryo resembles a flattened disc and consists of three layers of cells: the *mesoderm* (also known as the >primitive streak, from which, among other things, the skeleton, musculature, heart and blood system and kidneys will develop); the *endoderm* (digestive system, lungs, liver and other glands); and the *ectoderm* (brain and nervous system, skin, hair, etc.).

At 20 days, the embryo, now with a definite head end, is about 1.5 mm long and begins to acquire a segmented appearance; the segments are called *somites*. The brain and spinal cord begin to be formed (and are in place by 30 days); at 25 days, the lungs and the gut canal are formed and differentiated; and at 28 days, the heart is functioning and the sense organs are formed. The embryo is now about 5 mm long. The limb buds (arms first) appear at about 30 days, when the embryo is about 7 mm long. The internal sex organs develop from about 25 days, and by 45 days, the external genitalia are visible.

At the end of the second month (i.e. at 60 days), the embryo is about 35 mm ($1\frac{2}{5}$ in.) long and weighs about 1 gram. From this point, the individual is called a >foetus. All its essential features are present, and it grows very fast: about 1.5 mm per day in length, and increasing its weight so that, at birth (about 266 days from fertilization), it weighs about 3300 times what it did at 60 days.

Emergent evolution, sudden evolutionary change through rearrangement of existing genetic elements. *See also* saltation.

Endogamy, a cross, or a pattern of mating within a population, in which the parents are closely related. The opposite is >exogamy. *See also* inbreeding.

Endomitosis, process in which chromosomes replicate but remain within the same cell nucleus. One example of this is in the giant >polytene chromosomes in *Drosophila*: the chromosomes replicate dozens of times but the new copies are kept lying exactly alongside the old so that a sort of chromosome ribbon results.

Endonuclease, an enzyme that cuts a molecule of DNA or RNA som
where in the middle. Some DNA endonucleases cut both strands; oth
make a nick, i.e. cut one strand only. RNA is usually single-stran

>Restriction endonucleases cut only at specific sequences. *See also* exonuclease.

Endoplasmic reticulum, a net-like structure in the cytoplasm within a cell which has ribosomes embedded in it. It is thought to be the site of protein synthesis and modification. Newly synthesized proteins that are to be exported from the cell are pinched off the endoplasmic reticulum and move to the cell membrane from where they are released.

End-product inhibition. *See* feedback inhibition.

Enhancer element, a DNA sequence in eukaryotes which increases the transcription rate of genes near it. The element does not have to be exactly adjacent to the genes whose activity it is enhancing, and the sequence can be facing either way. Enhancer elements can be specific both to species and to tissue. For example, the enhancer elements associated with the transcription of immunoglobulin genes will only function in lymphoid tissue.

Environment, in genetics, anything that influences the phenotype of the individual, apart from its genotype (*see* nature versus nurture). It is also of course the setting in which the phenotype must function, and in which natural selection operates.

Environmentalism, the belief that the environment is more important than the genotype in determining human traits of social importance. The opposite is hereditarianism. *See* nature versus nurture.

Enzyme, any protein whose function is to catalyse a particular chemical reaction while remaining unaltered itself. Enzymes work by means of their three-dimensional shape, with sites that bind the molecules (referred to as >substrates) that will be involved in the reaction. Their names end in *-ase*, and they are called after their substrate and/or the type of reaction that they initiate, e.g. adenosine deaminase. It is a tenet of genetics that each enzyme is coded for by a single gene. *See* Beadle, G. W.; Garrod, Sir Archibald; Tatum, E. L.

Epigenesis, the idea that the development of the organism from the fertilized egg consists of ever greater elaboration of structures and specialization of tissues. The opposite is (or was, for epigenesis is now totally accepted and has ousted its rival) >preformation.

Epigenetic landscape, an idea put forward in the 1950s by the Scottish geneticist C. H. Waddington. He proposed that the fate of a cell during embryogenesis is like the journey of a pebble down a hill: for most of the time it just rolls down the valley in front of it, but at times the valley forks nd the pebble must choose one way or the other – a small choice at the ne but one that will perhaps have a great influence on where the pebble s up. *See also* von Baer's law.

Epilepsy, a condition of variable severity, in which the individual suffers from intermittent attacks of convulsions. Most cases of epilepsy develop during childhood (70 per cent). Epilepsy has a genetic basis, although the mode of inheritance is not simple and there are also environmental factors involved. In European populations about 5 people out of every 1000 suffer from epilepsy.

Epistasis, the situation in which alleles at one locus have an effect on the expression of alleles at another locus. One classic example is the shape of the comb in hens, in which there are at least two alleles at each of two loci. The shape given by genotypes *pp*, *Pp* and *PP* at one locus depends on whether or not the dominant allele *R* is present at the other locus.

Epstein-Barr virus. *See* herpesviruses.

Equilibrium, the situation in which gene frequencies do not change from one generation to the next.

Error catastrophe, a theory that attempts to explain ageing, proposing that there is positive feedback in the mechanism of the DNA that codes for RNA molecules concerned with DNA transcription, such that beyond a certain point transcription breaks down completely and the cell dies. This process, repeated in numerous cells in the body, could give the effects of ageing.

Escherichia coli (abbreviation: *E. coli*), a bacterium found in the large intestine of humans and many other mammals. It is normally harmless, but can cause trouble if it gets into the bloodstream, the kidneys or the bladder (cystitis). An immense amount of basic research in genetics and biochemistry is done with *E. coli*, partly because it is a >prokaryote and therefore its genetic machinery is simple. It is also a very convenient organism, as it reproduces quickly, is easy to keep and, if handled properly, is non-pathogenic (i.e. does not cause disease).

There are tens of thousands of different strains, and the ones most often used in the laboratory have become so 'domesticated' (by living on prepared nutrient media) that they are unable to survive in the human gut. They are also deliberately mutated so that they cannot survive in the 'wild'. This minimizes risks arising from escapes of bacteria that can carry recombinant DNA molecules.

Some of the major discoveries that have been made in *E. coli* experiments include conjugation in bacteria (Joshua >Lederberg and E. L. >Tatum, 1946), semi-conservative replication of DNA (Matthew >Meselson and F. W. Stahl, 1957) and the operon (François >Jacob and Jacques >Monod, 1961). *E. coli* is one of the workhorses of genetic engineering.

Ethology, the study of the behaviour of animals in the wild. The main concepts are instinct and learned responses, both of which have genetic and environmental input. The subject was pioneered by Konrad >Lorenz and Nikolaas Tinbergen.

Euchromatin, chromatin (the substance that chromosomes are made of) that is loosely wound because the DNA is being actively transcribed, and is accessible to >mRNA in that state. Euchromatin shows up more lightly under the microscope than the condensed >heterochromatin. *See also* banding.

Eugenics, the belief in or policy of attempting to improve the genetic quality of a human population. 'Positive eugenics' is the encouragement of those believed to have good genotypes to have more children. 'Negative eugenics', far more widespread, aims at the eradication of 'bad' genes, either by discouraging persons with presumed inferior genotypes from breeding or by eliminating such persons altogether. Both are surprisingly ineffective from the point of view of removing undesirable genes from the human gene pool. In the case of dominant mutations, these are usually so severe that the affected people do not reproduce at all or only infrequently; far the greatest number of persons with such conditions represent new mutations, and eugenics can do nothing to alter the rate at which these will occur. In the case of recessive mutations, the frequency of heterozygous carriers is far higher than the number of persons with the homozygous condition, and removing the two genes together each time they occur in one individual has virtually no effect on the gene's frequency in the whole gene pool. >Genetic counselling is not, strictly speaking, a manifestation of eugenics, in that it deals with the breeding policy of couples, not of the population as a whole. *See also* pp. 40–3.

Eukaryote, any organism that has within its cell or cells a membrane-bound nucleus containing chromosomes and all the rest of the genetic apparatus. All creatures from Protoctista upwards (*see* classification) are eukaryotes; the earlier, simpler type is the >prokaryote.

Euphenics, the policy of altering the environment so that the deleterious effect of a particular gene is not expressed. The word was coined by the American geneticist Joshua >Lederberg. There are several examples among the inborn errors of metabolism. For example, babies born with phenylketonuria grow up perfectly normally if their environment is controlled, i.e. if their diet is kept low in phenylalanine. This is the converse of the approach of >eugenics, the policy of which is to remove undesirable genes from the human gene pool. Indeed eugenicists and others object to euphenics on the grounds that it deliberately maintains unfavourable genes. This need not always be the case, one counter-example being sickle-cell anaemia: if the environment is altered by eradicating malaria, the gene for sickling loses its selective advantage and becomes rarer.

Euploidy, the condition in which the number of chromosomes in the cells of a >polyploid species is a precise multiple of the diploid number of the original non-polyploid species.

Evolution, the process by which populations of organisms change their genetic constitution through time. During this process, different populations become unlike enough to each other or to the form that preceded them to be thought of as new species. (Note that evolution is something that happens to a population, which is a collection of individuals, not to the individuals themselves.)

Within any species, new variant forms are constantly being produced by mutation, but in very low frequencies; as they are caused by genetic mutation, the changes shown by these variant forms are inherited. If the environment stays constant, natural selection will operate to eliminate these variants and to keep the species to its present type. But if the environment changes or if a new ecological niche suddenly presents itself, variants may be better able to reproduce offspring than the original type, and thus their numbers will increase proportionately in the next generation. Over many generations, the variant form entirely replaces the original type, so that a new species has evolved. What selection 'chooses' is not the ability to survive or not, but the ability to produce viable offspring; this is known as >fitness. There are a number of pieces of evidence for evolution:

- The principle of >succession shows that, during geological time, one form has given way to another, and that forms have become more complex over time.
- The principle of >distribution shows that more closely related forms tend to be geographically near each other.
- Mutations are constantly seen to appear in natural populations.
- New species have arisen in the few centuries of observation carried out by biologists.

The theory of evolution by natural selection is accepted throughout the scientific world, but its old rival >creationism still survives in a few places.

See also acquired characteristics, inheritance of; natural selection. For the history of the idea to its formulation in 1859 by >Wallace and >Darwin, *see* pp. 15–18.

Evolutionary rate, the rate at which the genetic characteristics of a species change. In times of rapid environmental change, it can be very fast, or in a constant environment, it can be zero. A species that is perfectly adapted to its environment does not evolve at all, though this is a very rare case, as evolutionary change in one species has an impact on the other species within its range. Human evolution has shown a very fast evolutionary rate, believed to be partly due to the interaction between genetic and >exogenetic inheritance. Evolutionary rate is measured in a unit called a >darwin.

Evolutionary stable strategy (abbreviation: ESS), an idea, derived from >game theory, that suggests that a population may contain a number of different types, but only in ratios such that selection does not favour one over the other. The idea was developed by the British mathematical biologist John Maynard Smith, and one of his examples best illustrates it. Let there be two types, 'hawks' and 'doves'. When 'hawks' meet, they fight to the death; when 'doves' meet, they perform a lengthy but ritualized fight; and when 'hawks' meets 'doves', the latter retreat. It is not an ESS for the population to consist entirely of 'doves' because a 'hawk' mutant would have an immediate advantage, but neither is a population that is all 'hawks' an ESS because 'dove' mutants would spread (they have the advantage of not killing each other). Smith made calculations for which he assigned the following values: +50 for a win, 0 for losing, −100 for getting killed, −10 for wasting time in a ritualized fight. On these figures, the stable proportion (the ESS) worked out at 5 'doves' to 7 'hawks'. This happens entirely by natural selection acting upon the individuals; group selection is not involved.

Exobiology, the idea that there are living things on other planets than Earth. Philosophical speculation on the subject goes back to ancient times (Christian opinion settled down to the idea that the Almighty could have made other living creations but probably did not make human beings elsewhere). Present scientific opinion is that, although the findings of space probes so far rule out the existence of life (other than simple micro-organisms) on any other planets in our solar system, there is a high probability that the right conditions (of temperature, atmospheric pressure, water, oxygen, salinity and acidity) are met by many thousands of planets in our galaxy. There is no reason to assume that advanced life cannot have evolved on one or many of these; but there is no direct evidence that it has. *See also* origin of life.

Exogamy, a cross, or a pattern of mating within a population, in which the parents are unrelated. The opposite is >endogamy.

Exogenetic inheritance, inheritance by means other than via genes. Although cultural evolution in a general sense had been discussed for some time (Herbert Spencer wrote at length about it), the idea of exogenetic inheritance as being precisely parallel to genetic inheritance was first proposed in 1932 by T. H. Morgan, the pioneer experimental geneticist:

> There are, then, in man two processes of inheritance: one through the physical continuity of the germ-cells; and the other through the transmission of the experiences of one generation to the next by means of example and by spoken and written language. It is his ability to communicate with his fellows and train his offspring that has probably been the chief agency in the rapid social evolution of man.
> [Quoted in Jean and Peter Medawar, *Aristotle to Zoos: A Philosophical Dictionary of Biology*, London, 1984.]

The same idea has been more recently developed by Richard Dawkins, who has proposed the term >meme for a unit of exogenetic inheritance parallel to a gene. Dawkins goes further, saying that 'We biologists have assimilated the idea of genetic evolution so deeply that we tend to forget that it is only one of many possible kinds of evolution.' The theory is far from universally accepted.

Exon, a sequence of DNA that codes for part of a >polypeptide chain (part of a protein), in contrast to an >intron, which is a non-coding sequence intervening between exons. The exon–intron arrangement is a characteristic of eukaryotes. *See also* cistron.

Exonuclease, an enzyme that removes bases one at a time from the end of a molecule of DNA or RNA. *See* endonuclease.

Expressivity, the degree to which the trait determined by a gene is expressed in one individual or another. The term usually applies to genes that cause >continuous variation. An example is the recessive allele 'eyeless' in >*Drosophila*, which in some individuals causes only a slight reduction in the size of the compound eye and in others results in complete absence of eyes. There may also be a marked difference between the extent to which the two sides of the body are affected. Variations in expressivity are caused either by environmental factors or by other genes in the genotype. *See also* penetrance.

Extinction, the disappearance of a species. This can happen *(a)* if all the individuals are killed in a catastrophe; *(b)* if the habitat or some element in it is destroyed; *(c)* if reproduction fails for some reason; or *(d)* if population density drops so low that the few survivors have an insufficient supply of potential mates. The term is properly only applied to species, but is sometimes used to refer to the loss of a particular local population. Estimates of the rate of extinction of species are hard to make, as the *total* number of species is not known (*see* species, for the numbers that have so far been discovered). One estimate, based on a total of 5–10 million species, was that about one per day becomes extinct. Recently E. O. >Wilson has estimated that, if there are about 30 million species, the annual losses may be as high as 17,500 per year – which is nearly 50 per day. It is apparently considered morally acceptable to bring about the extinction of organisms that are harmful to humans.

Eye colour is usually taken as the classic example of Mendelian inheritance in humans. It is true that the allele for brown eyes is dominant over the allele for blue: two brown-eyed parents, genotypes *Bb* and *Bb*, could have a child with blue eyes, genotype *bb*, whereas two blue-eyed parents, genotypes *bb* and *bb*, could not have a brown-eyed child, genotype *Bb*. There is, however, more to eye colour than mere brownness or blueness, and the precise inheritance of the various shades of grey, hazel and so on is not so clear-cut; there seems to be at least one other locus involved, with alleles interacting with the main blue/brown one. Although in many ethnic groups black hair and brown eyes go together, the genes are actually independent, as can be seen in the 'Irish' colouring of blue eyes with black hair.

F

F_1, the first filial generation. This is the hybrid generation resulting from the cross between two pure strains. If one strain is homozygous *ZZ* and the other is homozygous *zz*, the F_1 genotype is the heterozygous *Zz*. The term F_1 can often be seen on seed packets, indicating that the seeds are hybrids; this is a good sign in that the resulting plants will have >hybrid vigour, but it also means that the gardener is not advised to keep seeds from these plants as they will not breed true (*see* F_2). *See also* hybrid.

F_2, the second filial generation. This is the generation obtained from crossing (or self-fertilizing) individuals of the >F_1 generation. If the F_1 individuals have genotype *Zz*, the individuals in the F_2 generation will have genotypes *ZZ*, *Zz* and *zz* in the ratio 1:2:1 (*see* pp. 19–20). The table (p. 114) shows how quickly the number of possible genotypes in the F_2 goes up as more gene pairs are added. *See also* hybrid; dihybrid; trihybrid.

Family tree, the type of diagram that genealogists draw. It shows 'ego' (Latin for 'I', the person whose family is being described) at the bottom, with parents in the row above, grandparents in the next row and so on, up through the generations. The same sort of thing that is used in medicine for diagnosis and genetic counselling is called a >pedigree. An evolutionary 'family tree' is a >dendrogram.

Fate map, a map of the cells or tissues in the fertilized egg or early embryo that shows where they will end up in the adult organism. The data can be collected by staining specific cells in the embryo with a non-diffusible dye; artificial >chimaeras can also be used.

Favism. *See* glucose-6-phosphate dehydrogenase deficiency.

Table showing relationship between the number of gene pairs and the F_2 progeny's genotypes and phenotypes (assuming dominance)

Number of gene pairs	Number of different types of gametes produced by F_1	Number of possible combinations, i.e. genotypes of the F_2 progeny	Number of phenotypes in F_2
1	2	3	2
2	4	9	4
3	8	27	8
4	16	81	16
5	32	243	32
6	64	729	64
7	128	2187	128
8	256	6561	256
9	512	19683	512
10	1024	59043	1024
n	2^n	3^n	2^n

Fecundity, total reproductive capacity, as measured by the number of gametes produced by the individual (or population). It is usually greater than >fertility, which is the measure of actual reproductive performance.

Feedback inhibition, or end-product inhibition, a control system in which the product (e.g. an enzyme) inhibits further production of itself. Generally, the final product in a sequential line of gene products inhibits further synthesis of the first product.

Ferns. *See* pteridophytes.

Fertility, actual reproductive capacity, as measured by the number of viable offspring produced by an individual (or population). It is usually less than >fecundity, which is the measure of potential reproductive capacity.

Fertility drug, a drug given to a female to stimulate the ovary to produce ripe eggs. A compound that stimulates the body's own production of the hormone follicle-stimulating hormone (FSH) is used for this, both in cattle to produce multiple embryos for >embryo transfer, and in humans in the course of treatment for infertility, when it is often responsible for the woman becoming pregnant with more than one baby. *See* multiple births; quadruplets; triplets.

Fertilization, the creation of a new individual resulting from the fusion of two gametes. In animals, the gametes are the ovum and the sperm. When one sperm has penetrated the ovum's membrane, there is an immediate change in the membrane's chemistry so that other sperms are repelled (however, *see* polyspermy). Then the nucleus (strictly, the pronucleus, as it is haploid) of the one sperm arrives at and joins with the pronucleus of the ovum; the result is one diploid cell, which is the zygote. It is also the very beginning of the new and genetically unique individual. Protein synthesis begins immediately, proceeding to cell division and the growth of the embryo.

Fibrocystic disease. *See* cystic fibrosis.

Finger prints. Both these and the prints on the palms of the hands and the soles of the feet are almost entirely genetically determined, and no two people, not even identical twins, have exactly the same finger prints. The first person to study finger prints systematically was Galton, who saw that the key to the patterns was the triangular 'partings' that he termed *triradii*. He classified the patterns on the finger tips as *(a)* arches (no triradius), *(b)* loop (one triradius), and *(c)* whorl (two triradii).

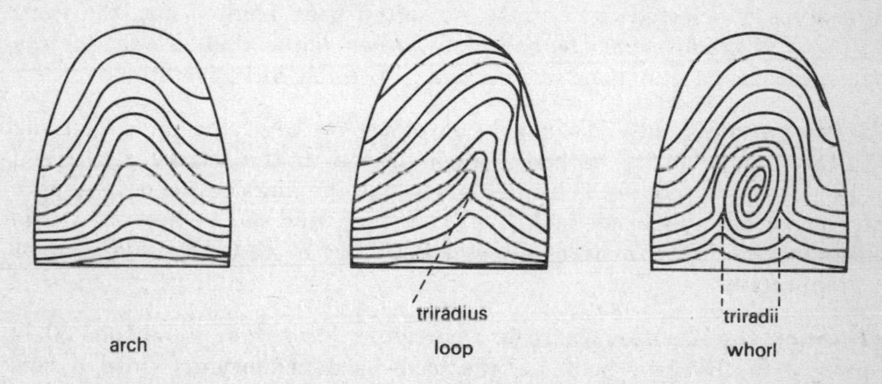

An individual's finger prints (and also their palm patterns) can be described not only in terms of these patterns but also the number of ridges between the centre of the pattern and its triradii. The frequency of patterns differs between populations. In Britain, loops are the commonest (70%), then whorls (25%), then arches (5%), but Bushmen have more arches and North American Indians more whorls than Europeans.

There is no theory as to what selective advantage one pattern might have had over another in prehistoric times. Although there are strong correlations between the patterns within families, neither a simple mode of inheritance nor a multifactorial one gives a good explanation.

Fisher, Sir Ronald Aylmer (1890–1962), British mathematician and geneticist. Fisher's first work in genetics showed that quantitative variation (e.g. height) in humans must be due to a small number of Mendelian genes acting together; this was the first proof that quantitative traits are controlled by genes and not (as was thought at the time) by some 'blending' process. This work led to Fisher's appointment in 1919 as the first-ever statistician at the Ministry of Agriculture's Rothamsted Experimental Station, where he found over 60 years of results of agricultural field trials awaiting analysis. He laid the foundations for all subsequent applications of statistics to experiments in the life sciences. His most important work was *The Genetical Theory of Natural Selection* (1930), which proved that Darwinian evolution was not only compatible with Mendelian inheritance but depended on it as the underlying mechanism; this had been much debated, but Fisher's mathematical proofs settled the matter. A convinced believer in eugenics, Fisher had eight children as a matter of principle.

Fitness, the relative ability of one >genotype to reproduce itself, compared to the reproductive ability of other genotypes. Fitness is a relative quantity: the fittest genotype in the population has a fitness of 1; a genotype with no fitness at all (because it is lethal or infertile) has 0; others are somewhere in between. These values show to what extent a particular genotype is favoured by >natural selection. Fitness can also be expressed as $1 - s$, where s is the >selection coefficient, a measure of how much the genotype is selected against. As has often been pointed out, the word 'fitness' is troublesome, as are many other words that are taken into scientific usage from general vocabulary. *See also* inclusive fitness.

5′ ('five prime'), one end of a molecule of RNA or DNA, the name referring to the number of the carbon atom in the sugar that comes at the end (joined to a phosphate). The other end is >3′, the links between the sugars throughout the molecule being between the 5′ and the 3′ carbons. A code sequence of DNA written out starts at the 5′ end. (*See* diagram at >replication.)

Fixation, the situation where for a particular locus there is only one allele present in the gene pool, i.e. the locus is monomorphic. Until a new mutation arises, all individuals are homozygous for that allele.

Flowering plants, the higher forms of plants which reproduce by means of the sexual parts contained in the flowers. A flower's pollen, carried by the anthers which are at the end of the stamens, is equivalent to sperm in an animal. The female part is the pistil, which consists of the stigma, which receives the pollen grain, and the ovule (equivalent to the ovum in an animal) contained inside the ovary (which becomes the fruit). In many plants, anthers and ovules are present in the same flower, and in others (e.g. cucumbers), there are separate male flowers and female flowers on the same shoot; some plants (e.g. holly) have the two sexes on different plants. Insects, especially bees, are important in pollen transfer, but some species rely on the wind to carry it. Many species have evolved devices to prevent >self-fertilization. Sexual reproduction in plants was first described by >Camerarius. The flowering plants include virtually all the economically significant plants, notably all the grain crops and the non-coniferous trees and shrubs. *See also* gymnosperms; pteridophytes.

Fluctuation test, a test to determine whether mutations in bacteria occur at random or whether they are evoked in response to environmental factors. Using statistical analysis of the fluctuations in numbers of favourable mutations, the American geneticists Luria and Delbruck proved (in 1943) that mutations are random with respect to the environment.

Foetus (pl. foetuses), the unborn mammalian embryo from the time that it has the recognizable features of its final form. In humans this is from the end of the second month after conception. The word can also be applied to animals, though it is much more often used in connection with humans. *See* embryogenesis.

Footprinting, a technique for finding DNA sequences that have a protein bound to them which protects them from being cut by endonucleases. Two samples of the whole DNA molecule are prepared, one with the protein and one without; both are digested with a restriction endonuclease and the resulting fragments are run on an >electrophoretic gel. When the two gels are compared, the bands that have resisted cutting by the enzyme (because of being protected by the protein) will show up because they do not match the bands on the other gel. In this way, DNA sequences that bind proteins, such as regulatory sequences, promoters, and so on, can be identified.

Forward mutation, a change from the normal, or wild-type, version of an allele to a mutant form. The vast majority of mutations are of this type, the opposite being >back mutation.

Founder effect, the principle in population genetics that a small group of individuals colonizing a new habitat are unlikely to carry a representative sample of the gene pool of the population that they come from. The new population therefore has different gene frequencies from the start. Founder effect is believed to be an important factor in the formation of new species. For instance, the extraordinary diversity of the forms in the Galápagos Islands (*see* pp. 15–16) is probably due to a considerable extent to it. *See* quantum evolution.

Fragile X syndrome, a genetic disorder causing mental retardation in males. It is the result of a mutation in the region at the very end of the X chromosome, which is, for some unknown reason, particularly liable to lose pieces – hence the name 'fragile X'. Some female carriers are also slightly affected.

Frameshift mutation, a >mutation in which one or more base-pairs are inserted into or deleted from a gene, so that the >reading frame of every codon after that point is shifted along. For example, if the original code was UCU–CAA–AGG–UUA, and the mutation put an extra U at the beginning, the message would be read as UUC–UCA–AAG–GGU and so on.

Franklin, Rosalind (1920–58), British X-ray crystallographer. Franklin graduated in chemistry at Cambridge, and did war work on coal utilization. In 1951 she joined the Medical Research Council's Biophysics Unit at King's College, London, where she worked on the study of >DNA via its X-ray diffraction pattern. She discovered that DNA occurs in two forms, and suggested a helical structure for the form that occurs in the natural state. Her calculation of how much water this form contains was one of the clues that Francis >Crick and James >Watson used to find DNA's structure, and her X-ray photographs were even more important to them. Franklin's early death from cancer prevented her from sharing the 1962 Nobel prize with Crick, Watson and her colleague Maurice Wilkins. The portrait of her in Watson's *The Double Helix* is grossly unfair, as Watson later admitted.

Freemartin, the female of a pair of twins in cattle, sheep, goats and pigs, when the other twin is male. The placentas for the two twins become joined and allow sex hormones of the male twin to interfere with the development of the female twin's ovaries and other sexual parts. Thus a freemartin is hermaphrodite in appearance and is almost always sterile. The male twin is unaffected.

Fungi, one of the five kingdoms of living things (*see* classification), containing a number of genera that are economically important. The most important of all is yeast (*see Saccharomyces*). Much genetic research has been done on other fungi, including the moulds >*Aspergillus* and >*Neurospora*. Although they are plants, fungi are unable to photosynthesize (i.e. utilize energy from sunlight to synthesize organic compounds), so they have to obtain their food by living off other plants or animals, alive or dead. Fungi can reproduce either by cell division (equal or unequal, the latter including either budding or the production of spores) or by sexual reproduction, in which case they have >alternation of generations, with haploid and diploid forms.

Fusion gene, a gene resulting from the accidental mixing of parts of two different genes. This may happen as the result of a small deletion occurring between two genes, leaving their ends joined up. Fusion genes can also be made *in vitro* using recombinant DNA technology.

G

G_1, G_2. *See* cell cycle.

Galactosaemia, a genetically determined disease in which the liver degenerates from early infancy owing to the inability to metabolize the sugar galactose; there is also serious mental deterioration. If the condition is diagnosed at birth, the baby can be put on a galactose-free (i.e. non-milk) diet, in which case there are no ill effects. This diet can be relaxed after the age of five. The genetic cause is an autosomal recessive, and about 1 in 50,000 babies is affected. Heterozygotes (i.e. >carriers) among the siblings of an affected child can be identified by biochemical tests.

Galápagos Islands. *See* pp. 15–16; founder effect.

Galton, Sir Francis (1822–1911), British scientist. Galton, a cousin of Charles Darwin, was a scientist of wide-ranging interests. He stressed the importance of quantitative evidence, and was first a geographer and meteorologist (he discovered and named anticyclones) before turning to biology. Taking up Darwin's theme of evolutionary progress, he applied it to humans, saying that it stood to reason that the human species could be improved by selective breeding. His general idea, for which he coined the term >eugenics, was that 'better' people should breed more, and he cannot be held responsible for the later type of eugenics whose thrust was that 'worse' people should breed less (or not at all). He carried out many practical studies, the results of which included *Hereditary Genius* (1869) which attempted to show, in pre-Mendelian terms, that intellectual achievement is inherited, and proved that an impecunious peer who marries an heiress is highly likely to have got himself a barren wife (being the only child of subfertile parents, she may well be entirely infertile herself).

Galton was the first person to realize that the study of identical twins could give evidence on the >nature versus nurture question, and he collected much data in this field (*see also* twin studies). His system of classifying finger prints is still in use throughout the world today, but perhaps his greatest achievement was the calculation of the >correlation coefficient which is one of the basic tools of biometrical genetics.

Galton's law, principle stated by >Galton concerning the degree of genetic similarity between relatives. The same idea has been restated in post-Mendelian terms as the coefficient of >relationship.

Game theory, a mathematical way of assessing strategies in real-life situations, originally devised by the Hungarian–American mathematician John von Neumann in 1944. The real-life situation, whatever it may be (e.g. a military confrontation, an economic crisis), is likened to a game in which the players make choices between possible strategies, which are limited by the rules of the game. Each player is deemed to want to win, or at least to prevent an opponent winning; player A does not know what player B is going to do, and vice versa. By analysing the possible outcomes algebraically in terms of points won or lost, one can work out what strategy is statistically the most likely to be the winning one.

The idea of using game theory to analyse what happens in the evolution of species was proposed by the British geneticist C. H. Waddington in his book *The Strategy of the Genes* (1957). When used for this purpose, there is one substantial disadvantage: game theory is based on choice – i.e. upon purposive behaviour – but clearly there is no sense in saying that an animal or species 'chooses' one evolutionary strategy rather than another. In the simplest applications of game theory to evolution, one of the 'players' is the environment (and this also cannot be thought of as having any intentions). In other examples, the environment is considered to be a static background against which the 'players' compete with each other; sometimes the players are different genotypes within a species (*see* evolutionary stable strategy), sometimes species competing for a resource (*see* *K* selection; *r* selection; red queen hypothesis). *See also* mimicry.

Gamete, one of the reproductive cells; also called a germ cell. In animals they are the ovum and the sperm; in plants, the comparable cells (though not quite the same) are the ovule and the pollen grain. A gamete is haploid, so that when fertilization takes place, the newly formed individual (the zygote) is diploid.

Gametophyte, the >haploid part of a plant's life cycle. In higher plants it consists of only a few cells, which make up the pollen grain or embryo sac. *See* alternation of generations.

Gamma-globulin, a class of proteins, comprising the >antibody molecules in the immune system.

Garrod, Sir Archibald (1857–1936), British physician, pioneer of medical genetics. In 1903 he identified four inborn errors of metabolism – albinism, alkaptonuria, cystinuria (not in fact an enzyme deficiency) and pentosuria – and realized that they were examples of Mendelian inheritance in humans. He also saw them as evidence that one gene codes for one enzyme, a hypothesis ignored at the time and only taken up (by >Beadle and >Tatum) some 40 years afterwards, when it became the cornerstone of biochemical genetics.

Gaussian distribution. *See* normal distribution.

Gene, the unit of heredity. A gene is a sequence of >DNA, occupying its own place (*see* locus) on a chromosome. Most genes are structural, i.e. they code for a particular protein; these genes are divided into >housekeeping and >luxury genes. Other genes code for the RNA molecules that are necessary for protein synthesis, and others again provide recognition markers for the polymerase enzymes involved in >gene regulation. A gene may exist in two or more alternative forms, called >alleles. Genes do not change except in the very rare event of >mutation; there are many genes in the human genome that have been passed unchanged down the generations for millions of years, since before we became humans (*see* molecular evolution).

The word 'gene' was coined (in German, as *Gen*) by >Johannsen in 1909, quite a time after Mendel's republished work had identified the existence of such a thing. Before that, there had been an appreciable linguistic difficulty; for example, in the course of a single article, T. H. >Morgan used five different words for what is now called a gene.

Gene amplification, the process by which extra copies of a gene that is needed temporarily are made. This can occur naturally, as in >*Drosophila* females who multiply the genes that code for the proteins that compose the membranes of their eggs: all the eggs are matured within about five hours, and to achieve this, the relevant genes are multiplied about ten times. Gene amplification can also be artificially induced, and has been observed in mice, who respond to a sub-lethal dose of cadmium by multiplying copies of the genes coding for the protein metallothionein that neutralizes the poison. *See also* silk.

Gene cloning, the technique of taking a complete gene from an organism's genome and inserting it, via a plasmid, into a host bacterial cell. To make sure that the gene will be expressed by the bacterial cell, regulatory regions as well as the coding sequence of the gene must be included. This technique, perhaps the most basic one in genetic engineering, can be used to produce large amounts of the product of any gene that has been isolated. Human genes have been cloned in >*Escherichia coli* to produce once-scarce proteins for treatment of people whose own gene is defective; these proteins include growth hormone and insulin.

G

Gene divergence, the process by which a sequence of related genes arises from an original single gene that has been duplicated a number of times and then undergone divergent changes in the different copies.

Gene dosage, the number of times that a gene is represented in an organism's cells, or total genome. It can range from as low as one, if there is a single-copy gene on the X chromosome in animals, to many thousands for a highly duplicated gene, as exemplified by the genes coding for rRNA in amphibians.

Gene flow, the movement of genes from one population into a neighbouring one. It may happen by interbreeding or migration or both.

Gene frequency, the frequency with which a particular allele occurs at a given locus. It is expressed as a decimal: if there are only two alleles, the gene frequencies could be, for example, 0.1 and 0.9 (the total must add up to 1). (*See* Hardy–Weinberg equilibrium.) Evolution is really a matter of changing gene frequencies: the old allele gets rarer as the new one comes in, and finally the new one may become fixed (*see* fixation), so that there is a complete change in the genetic constitution of the population in respect of that locus. Taken over a sufficient number of loci (which could perhaps be rather few), this gives rise to a new species. Changes in gene frequency are brought about by >natural selection, >gene flow, >founder effect and >genetic drift.

Gene library, a collection of the genes from a species, maintained as gene clones. It is different from a >DNA library in that the items stored are whole genes, not DNA fragments.

Gene pool, the total genetic material present in a sexually reproducing population. To put it the other way round, a population is a group of individuals who share the same gene pool.

Gene product, the molecule that is made from the information in a single gene. In the majority of cases it is a protein (or polypeptide, which is just a small protein), but some genes code for a sequence of RNA. *See* protein synthesis.

Gene regulation, the various systems by which genes that are not constitutive (i.e. in use all the time) are switched on and off. Gene regulation is the means by which cells specialize into the very many different cell types found in even quite simple organisms, and by which organisms respond moment by moment to changes in their external and internal environment. Most of what is known about regulation has been discovered in prokaryotes, and eukaryote gene regulation is still hardly understood. The key feature in prokaryotes is the >operon, a sequence of DNA in which the regulatory genes and the several structural genes (e.g. for a metabolic process) are in a line, one next to the other. Some operons are inducible, that is they are not transcribed until they are switched on by the presence of some molecule within the cell. Others are repressible, i.e. they are normally transcribed until they are switched off by some molecule within the cell.

Gene switch, the moment when a developing organism changes over from one gene (or set of genes) to another related one. A classic example is the change in humans from foetal haemoglobin to the adult equivalent. The series of 'choices' made by gene switching are irreversible. *See* epigenetic landscape.

Gene therapy, two possible techniques for curing genetic disease. In *germ-line gene therapy*, a normal gene to replace the defective one would be implanted at a very early stage of embryogenesis, so that the new gene was incorporated into the individual's germ line (i.e. ova or sperm); this is generally considered unethical for humans, and it is not being developed for clinical use. However, some of the research on human embryos now being considered in Britain and elsewhere will be into techniques that could be used for germ-line gene therapy.

Somatic-cell gene therapy is different in that the new gene is inserted into developed tissue and so cannot be incorporated into the germ line (and is therefore not heritable). Suppose that the patient is homozygous for a recessive gene that causes an inability to produce a vital enzyme, and that the normal form of the gene has been found and isolated. The normal gene would be cloned so that many copies of the DNA sequence were available. Next, a sample of the cells that should make the enzyme is taken from easily accessible tissue in the patient (e.g. bone marrow), and cultured. The cloned DNA is added to the cell culture in conditions likely to result in the cells taking up the DNA, and some means is found of identifying those cells that have taken up the DNA and are making the enzyme. Then these cells can be put back into the patient who should be able to integrate them into their own tissue and sustain them as a functional cell line. Genetic disorders that could possibly be suitable for this approach include Lesch-Nyhan syndrome, haemophilia, sickle-cell anaemia, thalassaemia and severe combined immune deficiency. An attempt in 1980 to use gene therapy to cure two patients with thalassaemia was a failure, but much technical progress has been made since then and it is realistic to expect gene therapy to be in clinical use by the mid-1990s.

Generation time, the average time in a species between the birth of an individual and its production of offspring.

Genet, an organism raised clonally (*see* clone) from a single zygote. *See also* ortet; ramet.

Genetic code, the meaning of the set of triplets (or codons) of nitrogenous bases within a DNA or RNA molecule. Each codon codes for one of the 20 >amino acids from which proteins are made. However, since there are four bases arranged in threes, there are 64 (i.e. 4^3) possible combinations; therefore most of the amino acids have alternative codons, which nearly always involve an alternative base in the last place in the codon (*see* wobble). There is no 'punctuation' within the genetic code, but three codons act as 'stop' codes (*see* protein synthesis for how the 'start' signal works).

The table below is based on the code in >messenger RNA. It corresponds with the DNA code that precedes it and the transfer RNA that follows it by simple complementarity, e.g. the codon CCG here comes from the codon GGC in DNA and will call up the anticodon GGC in transfer RNA. Note that RNA has U (Uracil) where DNA has T (thymine).

SECOND LETTER

FIRST LETTER

	U		C		A		G		
U	UUU	Phe	UCU	Ser	UAU	Tyr	UGU	Cys	U
	UUC		UCC		UAC		UGC		C
	UUA	Leu	UCA		UAA	Stop	UGA	Stop	A
	UUG		UCG		UAG		UGG	Trp	G
C	CUU	Leu	CCU	Pro	CAU	His	CGU	Arg	U
	CUC		CCC		CAC		CGC		C
	CUA		CCA		CAA	Gln	CGA		A
	CUG		CCG		CAG		CGG		G
A	AUU	Ile	ACU	Thr	AAU	Asn	AGU	Ser	U
	AUC		ACC		AAC		AGC		C
	AUA		ACA		AAA	Lys	AGA	Arg	A
	AUG	Met	ACG		AAG		AGG		G
G	GUU	Val	GCU	Ala	GAU	Asp	GGU	Gly	U
	GUC		GCC		GAC		GGC		C
	GUA		GCA		GAA	Glu	GGA		A
	GUG		GCG		GAG		GGG		G

AMINO ACIDS

Ala = Alanine
Arg = Arginine
Asn = Asparagine
Asp = Aspartic acid
Cys = Cysteine
Gln = Glutamine
Glu = Glutamic acid
Gly = Glycine
His = Histidine
Ile = Isoleucine
Leu = Leucine
Lys = Lysine
Met = Methionine
Phe = Phenylalanine
Pro = Proline
Ser = Serine
Thr = Threonine
Trp = Tryptophan
Tyr = Tyrosinase
Val = Valine

The genetic code is universal, the meaning of the codons being the same in every organism from amoebas to zebras. The code was deciphered in the late 1950s and 1960s by, among others, >Brenner, >Crick, >Khorana and >Nirenberg.

Genetic counselling, the advice given to people about the risk of their having a baby with a genetic disorder. Except in a very few cases where whole populations are screened for a harmful gene and carriers identified, genetic counselling usually takes place when a couple have reason to believe that they are at risk, because they have already had one baby with a defect or because a relative is affected.

The genetic counsellor begins by studying the >pedigree of the family and identifying the nature of the genetic problem. In some cases, e.g. when there is a dominant >mutation in a relative of one partner but that partner is him/herself normal, the couple can be assured that they are at no risk of having an affected child. Equally, if both partners are themselves normal but have had one child with a dominant defect, they can be reassured that they will not have another such child, as this must have been a case of a fresh mutation. If, however, one of the partners has a dominant defect, they will be told that there is a 1 in 2 chance of them passing it on to any child.

Recessive defects are more complicated, as the gene must be inherited from both parents; if a relative of one partner has such a condition, the question is whether that partner is a carrier. Diagnostic techniques are now available to ascertain for many such traits whether or not a person is a carrier; if there is no such test for the trait, then a probability can be calculated from the pedigree. If a couple has had one child with a recessive defect, then it is clear that they are both carriers, and that the risk of their having another is 1 in 4. The options that such couples can take are: *(a)* take the chance; *(b)* start another pregnancy but have prenatal diagnosis; *(c)* artificial insemination by donor; *(d)* egg donation; *(e)* embryo donation; *(f)* adoption; *(g)* have no more children.

Genetic death, the death of a gene, in the sense that the individual who carried it failed to reproduce. This may have been because the individual itself died as a result of carrying the gene, but not necessarily. From the genetical point of view, it is the fact that fertility was reduced to zero that counts.

Genetic distance, a measure of the amount of relatedness of two individuals or populations, in terms of the probability of their possessing the same alleles at a locus or loci. *See* consanguinity.

Genetic drift, random fluctuation in the frequency of a gene. According to some theorists (known as 'neutralists'), genetic drift is very widespread and accounts for much of the polymorphism in natural populations. It becomes an important effect in small populations when drift can cause the loss of an allele from the gene pool even though it was a useful one. Genetic drift was first described by the American geneticist Sewall >Wright.

Genetic engineering, the group of techniques that involve altering the natural state of an organism's genome. The core activity is transferring DNA (either a whole gene or part of one) from one organism to another. (The DNA so produced is called >recombinant DNA, and the term 'recombinant DNA technology' is usually used by geneticists rather than 'genetic engineering'.) There are various ways of isolating DNA required for transfer. For example, it can be found by digesting with >restriction endonucleases and cloning, or by >polymerase chain reaction. In the case of eukaryote genes, the messenger RNA produced during the primary transcription process can be isolated, and reverse transcriptase used to make a corresponding DNA sequence.

The foreign DNA cannot simply be placed just as it is in the host cell, since it would not be expressed or replicated (gene expression being controlled by a position-specific mechanism); it must be carried in via a particular DNA molecule that already has the property of being able to get itself expressed in the host cell. Such a carrier is called the vector, and it is usually a >plasmid, a bacteriophage or a virus. The DNA to be inserted is joined to the DNA of the vector, either by a sticky-end join (when a restriction endonuclease has left complementary tails on each), by blunt-end ligation or by adding a poly-A tail to one and a poly-T tail to the other so that they anneal. In addition to carrying the required DNA sequence, the vector must confer some property on the host cell that can be detected, e.g. antibiotic resistance or a marker gene, so that the experimenter can identify which host cells have assimilated the vector.

A gene transferred from one prokaryote (e.g. a bacterium) to another will generally be synthesized without difficulty, but owing to the different gene regulation system of eukaryote cells there are problems with eukaryote genes that are to be expressed in a bacterium. These can be overcome by using a >shuttle vector or by attaching the DNA to a plasmid containing the *lac* operon.

The above techniques are all used when the end result is (*1*) production of particular RNA or protein molecules in large quantities, (*2*) creation of novel bacteria with commercially useful properties, or (*3*) synthesis of vaccines. An example of (*1*) is the production of human insulin by synthesis in *Escherichia coli*, and of (*2*) the modification of a marine bacterium so that it metabolizes petroleum and can be used to break up oil slicks. Progress with (*3*) is very encouraging, with vaccines against hepatitis B, herpes and Epstein–Barr virus having been successfully made, and an anti-AIDS vaccine is a real possibility.

At a different order of magnitude is the problem of incorporating permanent genetic change in eukaryotes – that is, getting the foreign DNA incorporated into the germ line at an early stage of >embryogenesis. With plants, the gametes are fairly accessible, and genes can be transferred – for instance, by means of a virus. There is also the advantage of being able to regenerate whole plants from single cells that have had recombinant DNA inserted *in vitro*. Gene transfer has been done with animals: micro-injection of plasmids carrying recombinant DNA into the pronucleus of recently fertilized mouse eggs works well in terms of getting the DNA into every cell of the mouse embryo's body, but this is far from guaranteeing that the gene will be expressed. Genes can be transferred into animal cells *in vitro* using viruses as vectors, but again the problem is getting the gene expressed. *See also* gene therapy and pp. 34–6.

Genetic fingerprinting, a technique which distinguishes the individuality of a person's DNA. Devised by the British geneticist Alec Jeffreys, genetic fingerprinting is the analysis of the >Alu sequence that is present in many repeats in the human genome. The number of repeats, their position in the genome and their exact sequence differs from one person to another. The repeats are digested with restriction endonucleases and identified by Southern blotting. The technique has been used in forensic science in two ways. First it can identify with certainty whether a sample of blood, saliva or semen came from a particular individual. Second, it can establish whether a child is or is not the offspring of an alleged mother or father, given that the genetic fingerprint of the other parent is available; this is used by the Home Office in Britain to establish the relationship of dependants of immigrants. Fingerprinting has also been carried out on dogs and horses, to establish the genuineness of their pedigrees.

Genetic load, the hidden burden on a population of deleterious recessive mutations. While these alleles are in the heterozygous state, they do no harm, but once they meet in the homozygous state, they damage the individual and therefore the population.

Genetic map, a diagram that shows the position on a chromosome of the various genes that are known to be on it. A. H. >Sturtevant devised the method of using the frequency of >crossing-over between two genes as a measure of how far apart they are on a chromosome: two genes exactly next to each other will very seldom have a cross-over between them, but genes at opposite ends of a chromosome will be separated by any cross-over along the length of the chromosome (*see* the diagram at crossing-over). The basic technique of mapping is therefore to make large numbers of crosses involving two or more loci at which there are distinguishable variant alleles, and to record the number of times that recombination of the original types has taken place. The frequency of the cross-overs is expressed as a percentage, and one percentage point is known as a >map unit. When three or more loci are mapped in this way, the map units add up as they should. For example, if locus A and locus B have a 7 per cent cross-over frequency between them, and locus B and locus C have 10 per cent, loci A and C have cross-overs in 17 per cent of instances. Another classical method of locating genes on chromosomes is >deletion mapping, while a newer technique is >restriction mapping. *See also* Morgan, Thomas Hunt.

Genetic marker. *See* marker gene.

Genetic screening. There are two different types of genetic screening. The first is the search within a well-defined population or community for >carriers of a deleterious recessive gene, in order to advise them of the risk of having an affected child (*see* genetic counselling). This can only be done if *(a)* the community can be identified as having a significantly high frequency of the mutant gene; *(b)* there is a simple and cheap test to identify carriers; *(c)* there is a prenatal test to be used for pregnancies at risk; and *(d)* the procedure is acceptable to the community. Given these conditions, genetic screening can be successful in reducing the number of babies born with a particular defect; one notable example was the screening of the Jewish communities in Washington D.C. and Baltimore, Maryland for Tay–Sachs disease (for the >thalassaemia screenings, *see* pp. 42–3). An example of a screening programme that went wrong happened when, in the 1970s, black people in some American states were screened for sickle-cell anaemia. This failed because the screening was not acceptable to the community (point *d* above), largely because it was related to repressive legislation, including restrictions on carriers getting married.

The second type of genetic screening is the genetic inspection of persons for employment, insurance assessment and other commercial transactions. This is more a problem for the future than for the present, as genetic factors affecting employment (e.g. susceptibilities to particular workplace hazards) are not, at present, understood. The ethical issue will have to be confronted, as to whether it is justifiable to exclude a person from a job or benefit because of his or her genotype. On the face of it, there does not seem to be much difference between genetic screening and ordinary medical screening, where the civil rights of the person not to be discriminated against has to be weighed against the civil rights of other people who may be put at risk by that person's medical condition.

Genetic variance, the proportion of the total variance of a trait that is due to genes. It may have several components, due to the additive effects of separate genes, or dominance, or interaction. The rest of the variance is environmentally produced. The relative contribution of genetic factors to variance is not a fixed feature of any trait (*see* heritability).

Geneticism, the belief that human characteristics are shaped wholly or largely by genes. *See* heritability; nature versus nurture; IQ; and discussion on pp. 39–45.

Genetics, the study of heredity and variation, which as a unified branch of biology dates back only to the rediscovery of Mendel's writings in 1900. Since the ability to reproduce is the criterion of what is or is not a living thing, genetics as the study of the machinery of reproduction is the central life science, and from it all the others are derived. The range of subjects that have a direct relationship to genetics is enormous, from the more obvious ones such as medicine or microbiology to the (equally directly related) ecology and politics.

Genocopy, a genetically determined character that mimics the appearance of another genetically determined character, though caused by an allele at a different locus. *See also* phenocopy.

Genome, the total amount of genetic information that an organism possesses; the sum of its genes. In eukaryotes, this means the entire set of genes in a haploid set of chromosomes. The genome of a virus may be made up of either >DNA (single or double stranded) or RNA; it can be as small as 3500 base-pairs and may contain as few as three genes.

The genome of a prokaryote (bacteria, etc.) is contained in a double-stranded DNA molecule (not strictly a chromosome), and contains from about 750,000 to 4,500,000 base-pairs, comprising up to about 1500 genes. This molecule is extremely compact: in *Escherichia coli*, the circular DNA is 13 μm long and is packed into a cell 1 μm × 2 μm.

The eukaryotes have even larger genomes, and in general, the higher up the evolutionary scale, the larger the genome (but with exceptions: some species of amphibians and fish have genomes nearly 40 times larger than the average for mammals). The size of the genome bears no simple relationship to the number of chromosomes: closely related species may have either a large number of small chromosomes or vice versa, for roughly the same-sized genome. Humans have a genome of about 2900 million base-pairs of DNA, which, if it were all in one DNA molecule, would be 1 metre long. (This is the haploid total; each of our body cells has twice this amount.) It would be physically possible for there to be as many as one million genes encoded in that much DNA, but in common with other eukaryotes, humans have many instances of repetitive sequences of DNA that do not seem to have any meaning, and the actual number of working genes is probably much lower. There are also many genes that are present in multiple copies. The Human Genome Project is described on pp. 36–8.

Genotype, the genetic characteristics of an organism. This may refer either to the total of all the genes or, more usually, to the alleles present at one or two >loci of interest. The genotype is contrasted to the >phenotype (the appearance and physical characteristics of the organism) which may or may not be a direct reflection of the genotype. For example, where one allele is dominant, the phenotypes of both genotypes *AA* and *Aa* are identical.

Germ cell. *See* gamete.

Germ plasm theory, an early idea of inheritance, foreshadowing chromosomal inheritance. Put forward in 1875 by the German biologist August

>Weismann, the theory states that the material of which the germ cells (sperm and ova) are made is self-perpetuating and is entirely separate from that of all the other cells in the body (which Weismann termed 'somatoplasm'), which are mortal. This is the idea so neatly summed up by Samuel Butler's aphorism 'A hen is only an egg's way of making another egg.' Weismann's experiments to support his theory included cutting the tails off mouse pups for many successive generations; their offspring, of course, still had tails. Although crude, these experiments established Weismann's theory among scientists, simultaneously disposing of the theories of the inheritance of acquired characteristics and preformation.

German measles. *See* rubella.

Gigantism. *See* growth hormone.

Gilbert, Walter (1932–), American molecular biologist; professor of molecular biology at Harvard University since 1968, having been professor of biophysics 1964–8; Gilbert is one of several important molecular biologists who began as physicists (Crick is another). In 1961 Jacob and Monod had proposed that gene activity is controlled by a hypothetical 'repressor' that switches the gene off when it is not needed. Gilbert was successful in isolating and identifying a repressor, which was difficult because each repressor is only present in a cell in minute quantities. Later he devised a method for analysing the sequence of single bases on a strand of DNA. He shared the Nobel prize for chemistry in 1980.

Globins, a group of proteins involved in the transport of oxygen in blood, muscles, etc. They are genetically well understood. Changes in their constitution during evolution have been studied (*see* molecular evolution), as has their role in the hereditary anaemias (*see* sickle-cell anaemia; thalassaemia). *See* haemoglobin.

Glucose-6-phosphate dehydrogenase deficiency (G6PD deficiency), a hereditary lack of an enzyme involved in the metabolism of carbohydrates. Inability to produce this enzyme is inherited as an X-linked recessive. Under normal circumstances, people with this gene show no ill effects, but when they eat broad beans (also known as fava beans), they suffer a severe bout of haemolytic anaemia (anaemia caused by the destruction of red blood cells).

The gene causing G6PD deficiency is found at frequencies of up to 30 per cent in countries around the Mediterranean (Sardinia, Greece, Israel), in parts of Africa (especially the Congo Basin) and in India, and it seems likely that its distribution is connected with that of malaria, because the trait gives some resistance to malarial infection (as do the genes for >sickle-cell anaemia and >thalassaemia).

Gonad (from the Greek *gone*, 'generation' or 'seed'), an organ in an animal that produces gametes. The female gonads are the ovaries, and the male the testes.

Gradualism, or phyletic gradualism, the idea that the formation of a new species in geological time is a gradual process, proceeding at a more or less steady speed. *See also* catastrophism; punctuated evolution.

Graft, the transfer by surgery of any tissue or organ from one position on or in the body to another (an *autograft*), or from one organism into another. (The word >transplant is now more usual where a whole organ is involved.) A *homograft* is a graft where the recipient and the donor are of the same species; a *heterograft* is one carried out between two different species. Any graft except an autograft or a graft between monozygotic twins is likely to be rejected by the recipient's immune system, unless treatment with >immunosuppressive drugs is given. *See also* graft-versus-host reaction.

Graft-versus-host reaction, the situation in which, after a graft has been made, the immune tissue within the graft produces antibody against the cells of the recipient. The only time when this is likely to occur is after a patient has been given a bone-marrow transplant; in other circumstances, there are not likely to be any immunologically competent cells within the graft.

Group selection, >natural selection that acts upon a group of organisms, rather than upon the individual, so as to produce an evolutionary change that benefits the group (or its successors). >Darwin's theory of evolution was based on individual selection, and he never accepted group selection, though Wallace, co-originator of the theory, did so. The idea of group selection was revived in 1962 by the British zoologist V. C. Wynne Edwards, who suggested that it could explain various aspects of population density in animals (including >altruism). After a brisk controversy, genetical opinion settled to the viewpoint that group selection cannot happen because the mathematics do not work out. The idea has since been replaced with >kin selection, also a contentious matter.

Growth, increase in size of an organism, either by an increase in cell size or (much more usually) by an increase in cell number. Most organisms have a species-typical growth rate and final size, both of which are, to some extent, genetically controlled (as is shown by the fact that they can be changed by artificial selection) but are also much influenced by environmental factors. Some fishes and reptiles grow all their lives.

Growth hormone, in animals, a protein hormone produced by the pituitary gland in the base of the skull. Either deficiency of it or excess production can be genetically determined, the former causing a form of dwarfism and the latter gigantism (excess growth in childhood) and acromegaly (enlargement of the jaw, hands and feet in adult life). The dwarfism can be treated with human growth hormone, which formerly had to be obtained from human cadavers but can now be made by gene cloning.

G

Guanine (abbreviation: G), one of the four nitrogenous bases which comprise the core of DNA and RNA. It forms pairs with cytosine (C). *See also* replication.

Gymnosperm, a plant that has its seeds exposed rather than encased in an ovary (as >angiosperm plants do). Gymnosperms, which include coniferous trees, are wind-pollinated. *See also* pteridophytes (ferns, etc.).

H

Haeckel's law. 'Ontogeny recapitulates phylogeny.' Haeckel's theory was that each organism, during its embryonic life (>ontogeny) literally goes through 'being' each one of its primitive ancestors (>phylogeny). If this were so it would imply that evolutionary change can only take the form of adding extra chapters to the life histories already written. It is however true that the early stages of embryogenesis are more similar between groups of animals than their later stages are (*see* von Baer's law). Ernst Haeckel (1834–1919) was professor of zoology at Jena; his law was stated in his *Natürliche Schöpfungsgeschichte* (1867, English translation *History of Creation*, 1902).

Haemoglobin, the molecule that carries oxygen on the red blood cells, consisting of a complex of the protein globin and the haem group which actually binds the oxygen; it gives the red blood cells their colour. Genetically determined defects in the haemoglobin genes are the cause of some anaemias. *See* sickle-cell anaemia; thalassaemia.

Haemolytic disease of the newborn, a kind of jaundice that happens when a second or subsequent >Rhesus-positive (Rh+) baby is born to a Rhesus-negative (Rh–) mother. There is no problem during the first pregnancy because the foetal and maternal blood cells are kept apart by the placenta. But at the time of delivery the placental barrier breaks down, and some of the baby's Rh+ red blood cells get into the mother's bloodstream. The mother, whose body identifies these Rh+ blood cells as 'foreign' molecules, then produces antibodies against Rh+ (this does not happen instantaneously, so the first baby is not damaged while being born). If the mother has a second Rh+ baby, all goes well during pregnancy because of the placenta separating the baby's blood from the mother's, until the very last minute when the placental barrier breaks down; at that moment, the mother's cells get into the baby's bloodstream and attack its red blood cells, causing severe haemolytic anaemia (a lack of red blood cells because of their destruction). The newborn baby will probably die unless given an immediate total-exchange blood transfusion. The same thing can happen with ABO incompatibility between baby and mother, but it is very rare.

Haemophilia, a disease in which the blood does not clot normally. The gene responsible is the best known of the >X-linked recessive genes, partly because it occurred in the pedigree of the British royal family. Haemophilia affects about 1 in 10,000 babies; as with all X-linked inheritance, boys who have the gene show the condition while girls who have one mutant gene and one normal are unaffected and are known as carriers (it would be possible for a girl to have haemophilia if her father was a haemophiliac and her mother was a carrier). Queen Victoria was a carrier, the mutant having probably arisen either in her mother or not much further back down the maternal line, as there were no earlier haemophiliac relatives. Victoria passed the gene on to at least three of her nine children: one was Leopold, Duke of Albany, who died of the disease at the age of 29: another was the Princess Beatrice, whose daughter Victoria Eugenie was also a carrier and took the gene into the royal house of Spain; and the third was Princess Alice of Hesse, whose daughter Alexandra, Czarina of Russia, was the mother of the haemophiliac Czarevitch Alexis. In the past, mothers known to be carriers have had the option of aborting all male foetuses, even though this meant aborting a normal individual in 50 per cent of cases, but now genetic markers on the X chromosome can be used in certain families to predict whether a male foetus of a carrier mother will be haemophiliac. Treatment is also now possible with the injection of the clotting factor (Factor VIII) prepared from donated blood; unfortunately, unless the blood is rigorously tested and treated, it carries a risk of infection with blood-borne diseases such as hepatitis and AIDS.

Hairpin loop, a loop formed within a molecule of RNA when matching adjacent sequences become bonded to each other. This can also happen within a DNA molecule, with sequences on a single strand becoming joined to each other.

Haldane, John Burdon Sanderson (1892–1964), British biologist. Haldane is one of the major figures in twentieth-century British biology, partly because of his wide-ranging interests and partly because of his determination to bring politics into science (a Marxist, Haldane contributed hundreds of articles on science to the *Daily Worker*). His father was the leading physiologist of the time, specializing in respiration, and that was Haldane's own first field; he used himself as an experimental animal (as he described in the 1927 paper 'On being one's own rabbit') in research that was later applied to saving lives in submarine disasters. As a biochemist, he gave the first proof that enzyme reactions in the body obey the laws of thermodynamics. Then, in the 1930s, he turned to genetics and studied the inheritance of genetic disorders in humans, publishing (in 1935), the first human linkage map, of the X chromosome.

Half-siblings (half-sibs), brothers, sisters or a brother and sister having one parent in common but not both. Human half-siblings are of some usefulness in nature versus nurture studies, in that they give a comparison with full siblings who are twice as closely related, while the environment they share is broadly similar (though not exactly similar, it being unlikely that half-siblings are treated as much alike as full siblings).

Handedness, the innate tendency to use one hand in preference to the other. About 6 per cent of people in European populations are left-handed. There is evidence that left-handedness is inherited through the mother (i.e. by a >maternal effect), though it is not known how this works; it is also true that older women are more likely to have left-handed children. Left-handedness is more frequent in twins, owing to their general tendency to mirror-imaging, and curiously this effect is seen in dizygotic (fraternal) twins as often as monozygotic (identical) twins. Other species often show a favouring of one hand (or foot), but in all of them, this is randomly the right or left, with humans alone having a species-wide overall preference. Also inherited in complex fashion are two other behavioural asymmetries: the preferences as to which thumb goes on top in clasping the hands together and which arm goes on top when folding the arms. These are not related to each other or to handedness.

Haploid, adjective (from the Greek *haploos*, 'single') used to describe any cell or organism that has only one copy of each chromosome. The gametes (ovum and sperm) are haploid, in contrast to all the other cells of the organism, which have two copies of each chromosome and are called >diploid. Some algae, yeasts and fungi are haploid for the greater part of their life cycles. The haploid number of a species' chromosomes is referred to as n.

Haplotype, a set of alleles at very closely linked loci, coding for related products. The term is used especially for the HLA alleles of the human major histocompatibility complex.

Hardy–Weinberg equilibrium, a theorem which forms the basis of population genetics. It states that, for a two-allele locus, if the gene frequency of one allele is p, the frequency of the other is $1 - p$, which is called q. The frequency of the three possible genotypes is $p^2 + 2pq + q^2$. If the population breeds with random mating, these proportions will be maintained in succeeding generations, other things being equal (i.e. with none of the factors than can cause changes in >gene frequency in a population). The proportions are not affected by >recombination. The Hardy–Weinberg formula can be used to work out gene frequencies from observed phenotypes: if the recessive homozygotes (q^2) are 1 in 100 of the population, i.e. 0.01, then q, the gene frequency of the recessive allele, is the square root of 0.01, i.e. 0.1 or 1 in 10.

Hayflick phenomenon. *See* tissue culture.

H

Heart disease, an imprecise term covering a number of disorders of the heart, some of which are genetically determined. Heart disease, in the sense of deformities of the heart, is found as one of the effects of quite a few genetic syndromes, including Down's syndrome. Heart disease in the more general sense of the various degenerative diseases of the heart and circulation that start in middle age also seems to have a genetic component, which interacts with such obvious external factors as diet, exercise and smoking. In particular, a genetic pattern has been discovered for heart disease resulting from too much cholesterol in the blood (*see* hypercholesterolaemia).

HeLa cells, a line of human cancer cells that has been maintained in culture since 1951 and has been much used in cancer research (*see* cell line). It is named after the patient from whom the cells were taken: Henrietta Lacks, who had cancer of the cervix.

Helix, a spiral shape. It occurs in nature as the three-dimensional structure of >DNA, which is a *double helix*. The *alpha helix*, described by Linus Pauling, is the secondary structure of many proteins, i.e. they are twisted threads that are subsequently folded into a particular three-dimensional shape.

Hemizygote (adj. 'hemizygous'; noun 'hemizygosity'; from the Greek *hemi*, 'half', and *zugon*, 'yoke'), an organism that has a gene present in only one dose (*see* gene dosage), usually in the case where the gene is on the sex chromosome and the individual is of the heterogametic sex (the one that has two different sex chromosomes, e.g. mammalian males). *See also* heterozygote; homozygote; X linkage.

Hereditarianism, the belief that the genotype has more effect than the environment in determining human traits of social importance. The opposite is environmentalism. *See* nature versus nurture.

Heredity, the process by which characteristics are transmitted from one generation to the next. The term is only useful if it is confined to genetic transmission; an alternative word that covers all modes of transmission of traits is 'inheritance' (e.g. cytoplasmic inheritance, cultural inheritance).

Heritability, the proportion of the variability of the >phenotype of a trait that is attributable to genetic factors. Heritability is not a fixed property of the genes concerned but depends in each situation on how much variation there is both in genotypes and in the environment. For example, in one population everybody is >homozygous for the gene for black hair, but nevertheless some children have reddish-brown hair as the result of malnutrition; in another population there is little malnutrition, but the population consists of people with every combination of four different alleles for hair colour, so there are many shades of hair. In the first case, heritability of hair colour is effectively 0 per cent, and in the second case, 100 per cent.

Most of the research on heritability has been done on plants and animals used in agriculture, because heritability gives a measure of how much improvement in a trait can be achieved by selective breeding. The agriculturalist starts with a heterogeneous population of pigs, say, in which growth rate is, at the outset, highly heritable; then as the genetically better pigs are selected and become a higher proportion of the pig population, environmental factors become relatively more important in the difference between individual pigs, and heritability goes down (*see* selection limit).

Hermaphrodite, an animal that has the reproductive organs of both sexes (the equivalent in plants is >monoecious). Animals that are normally hermaphroditic include flatworms, leeches, land snails, some fish and some flies. They are capable of self-fertilization, unless they are sequential hermaphrodites (i.e. they have first one set of sex organs, then the other). Hermaphrodites do not have sex chromosomes (*see* sex determination). In humans, hermaphroditism is a rare abnormality. The person has both ovarian and testicular tissue, which may occur as (*a*) one ovary plus one testis, (*b*) two ovotestes or (*c*) one ovotestis plus either an ovary or a testis; the external genitalia are ambiguous. The condition is the result of a single ovum being fertilized simultaneously by an X and a Y sperm, so that the person is a >chimaera of XX and XY cells. *See also* intersex.

Herpesviruses, a group of viruses that infect animals, including humans. Their genome is double-stranded DNA. *Herpes simplex* causes a number of conditions in humans, ranging from cold sores to blisters on the genitalia to encephalitis, an inflammation of the brain that can be fatal; *Herpes zoster* causes shingles and chicken pox. Epstein–Barr virus was the first to be proved to be associated with cancer in humans (*see* p. 34), being one element in the causation of >Burkitt's lymphoma.

Hershey, Alfred Day (1908–), American microbiologist. Following Avery's suggestion that DNA is the carrier of genetic information, Hershey and his colleague Martha Chase proved that this is so, in experiments on the DNA in bacteriophages (viruses that infect bacteria). Hershey shared (with Max Delbrück and S. E. Luria) the 1969 Nobel prize for physiology or medicine for his work on how viruses reproduce.

Hertwig, Oskar (1849–1922), German embryologist; professor of zoology at Munich for 40 years. In 1879 Hertwig made the first observation of fertilization, when he saw an egg being fertilized by a single sperm (the species in question was a sea urchin). The fact that only one sperm was involved was crucial to the understanding that the sexes make equal contributions to heredity. Hertwig was also the first to understand the significance of meiosis. His brother Richard was also a distinguished zoologist, and they wrote a number of joint publications.

Heterochromatin, >chromatin (the substance of which chromosomes are made) that is in a highly condensed state because the DNA, not being actively transcribed, can be packed densely around the >histones. Examples are the DNA in the >centromeres and in >Barr bodies. Because of its more condensed state, heterochromatin is more intensely coloured when stained than ordinary chromatin (>euchromatin).

Heteroduplex, a >nucleic acid molecule that consists of two strands of different genetic origins. It can be either a double-stranded DNA molecule or a DNA–RNA combination. Where the sequences are homologous, the two strands will link; non-homologous sequences can then be seen under the electron microscope looking like bubbles. Examination of a heteroduplex made from DNA strands from wild-type and mutant-type DNA will show where there are deletions or insertions; this is called 'heteroduplex mapping'.

Heterogametic, adjective used to describe the sex that has the non-matching pair of sex chromosomes, X and Y. This sex therefore produces two different types of gametes, one with the X and one with the Y, and is responsible for the sex of the offspring. In mammals and most insects, the heterogametic sex is the male, but in birds, reptiles and some amphibia and insects, it is the female. Fish are variable, and some species even have different strains, some with heterogametic males, others with heterogametic females. The opposite is >homogametic.

Heterogeneous nuclear RNA (abbreviation: hnRNA), the >RNA molecule that reads the DNA code in eukaryotes. *See* messenger RNA.

Heterograft. *See* graft.

Heterokaryon, a cell or individual that contains two nuclei, each of different genetic origin. It can happen naturally when cells of two different fungi fuse, and it can be artificially induced by mixing the cells of higher organisms *in vitro*.

Heterosis, the technical term for the phenomenon more generally known as >hybrid vigour.

Heterozygote (adj. 'heterozygous'; noun 'heterozygosity'; from the Greek *hetero*, 'other', and *zugon*, 'yoke'), an organism that, at a given locus, has two unlike alleles. The term 'double heterozygote' refers to an individual who, at *two* given loci, has unlike alleles. Heterozygotes do not breed true with respect of the trait(s) in question. *See also* hemizygote; homozygote; F_1.

Heterozygous advantage, the situation where an organism that is heterozygous at a given locus has greater genetic fitness than either of the two homozygotes. It is one way in which two different alleles can be maintained within a population at stable frequencies: neither is better on its own, but the two in combination are the most successful genotype, and therefore neither allele is lost from the population (*see* polymorphism). The classic example of this is >sickle-cell anaemia in humans, in which homozygotes for the mutant allele have severe anaemia, homozygotes for the normal allele are liable to get malaria, but heterozygotes are neither anaemic nor subject to malaria. In a more general sense, organisms that are heterozygous at many loci are at an advantage over more homozygous ones; this is known as >hybrid vigour.

Himalayan rabbit, a rabbit that has a white body and dark-coloured ears, nose and feet. This coloration is an example of an environmentally caused gene switch: all the hair cells in the rabbit's skin have the same genes, but the pigment-forming enzyme that they code for is heat-sensitive and will only work in cooler tissues. Thus the gene is not expressed in warmer parts of the body, but only in the extremities. This first happens during the first week after the rabbit's birth, when the fur is growing, but the gene can be switched on or off in later life: if part of the skin is shaved, and the rabbit is kept in a cold place, the new fur growing in will be dark even though that patch was white before.

Hinny, the offspring of a male horse or pony and a female ass or donkey (the reverse is a >mule). The diploid chromosome number in the horse or pony (*Equus caballus*) is 64, and in the ass or donkey (*Equus asinus*), it is 62, so the hinny has 63 chromosomes, and is infertile.

Hippocrates, *c.* 460–*c.* 357 BC, Greek physician, 'the father of medicine'. Hippocrates taught at Kos, and was both famous in his lifetime and influential thereafter. His medical ideas were based on observation (his are the first real case-notes) and reason, a huge advance on the superstitions of the time. His ideas on genetics were also ahead of their time. Hippocrates knew that the male contribution to heredity was in the semen and he believed that there must be similar fluid in the female. Thus the contributions of the parents were equal, and what kind of baby was born was the result of a blending of the two fluids. Though not correct, this theory was much closer to the truth than >Aristotle's (*see* pp. 4–6). Not everything that is in the writings known as the *Hippocratic Collection* can be Hippocrates' own work, as they cover a span of over 100 years, nor can the Hippocratic Oath be credited to him, though it does reflect his principles.

Histocompatibility complex, the system whereby the body recognizes as foreign any tissue other than its own. It is medically important in that differences between donor and recipient lead to the rejection of >transplants. Each of the individual's own cells has on its surface a number of specific proteins, the histocompatibility antigens (also known as >HLA, human leucocyte antigens), there of which are six different types, each with dozens of allelic variants. The number of possible combinations is many thousands of millions, and the probability of any two people (apart from identical twins) having exactly the same combination is correspondingly remote, except that, as the genes involved are very closely linked, there is good correspondence between near relatives. Mismatch in some of them is apparently acceptable to the recipient's immune system, otherwise transplantation would be impossible. The genes controlling the antigens are grouped together into what is termed the major histocompatibility complex, which in humans is located on chromosome 6.

Histones, the DNA-binding proteins that are present in chromosomes and form the core of the >nucleosomes. There are five types that occur in eukaryotes: H1, H2A, H2B, H3 and H4. The last two are the most evolutionarily conservative proteins known. *See* molecular evolution.

HIV, or human immunodeficiency virus, the virus responsible for >AIDS. It was formerly known as HTLV or LAV.

HLA system (human leucocyte antigen system), the group of antigens that are involved in tissue recognition, enabling the body to recognize its own tissue and reject foreign tissue (as in transplants and grafts). There are six loci (known as A, B, C, DR, DQ and DP) and at each there are large numbers of alleles, designated by numbers (e.g. DR4). The loci are found in a group on chromosome 6, and are closely linked. It has been found that some HLA types have an association with auto-immune diseases. For example, ankylosing spondylitis is 120 times more likely to occur in people with allele B27, and multiple sclerosis is associated with three different HLA types: A3, with a risk factor of ×24; and B7 and B18, both with a risk factor of ×12.

Holandric, occurring only in males, i.e. a gene that is on the Y chromosome.

Homeostasis (from the Greek *homoio*, 'equal', *stasis*, 'state'), the tendency of a physiological system to maintain itself in balance, whatever the changes in the internal or external environment. Homeostasis is an evolved mechanism, operated by the switching off and on of many genes. An example is internal body temperature, which has homeostatic control in mammals but not in reptiles, whose body temperature varies according to the environmental temperature.

Homeotic mutation, a mutation that replaces a normal structure with another structure that is also normal but is in the wrong place. For example, there is a mutation in *Drosophila* that causes legs to grow where the antennae should be.

Homo erectus. *See* human evolution.

Homo sapiens, the species to which all human beings alive belong. In the aftermath of the publication of Darwin's *On the Origin of Species* in

1859, there were attempts to show that not all humans were of the same species (*see* pp. 39–40), but the matter is settled by the fact of the ability of all human beings to interbreed (as stated in the eighteenth century by the philosopher Immanuel Kant, among others). We are classified as a member of the order Primates within the class Mammalia of the phylum Chordata in the kingdom Animalia.

Homo sapiens has 22 pairs of autosomal chromosomes plus two sex chromosomes (XX or XY); the amount of DNA in each human somatic cell is about 6 picograms, and if fully extended would be about 1.9 metres long. The number of genes is estimated at somewhere in the region of 100,000 (*see* discussion of Human Genome Project on pp. 36–8). Humans also have an extensive (but not unique) system of >exogenetic inheritance. *See also* human evolution.

Homogametic, adjective used to describe the sex that has the matching pair of sex chromosomes, i.e. two Xs. This sex can only produce one type of gamete, with a single X chromosome, and therefore does not determine the sex of the offspring. In humans (and all other mammals) the female is the homogametic sex. *See* heterogametic; Barr body.

Homograft. *See* graft.

Homologous chromosomes, the members of a pair of chromosomes that carry the equivalent set of genes. All the >autosomes come in homologous pairs. In the case of the sex chromosomes, the X and the Y are not homologous as they do not carry the same gene sequence (but two X chromosomes are homologous with each other). *See* meiosis.

Homology, the correspondence of structures between one species and another, even if they are no longer performing the same functions. Homologous structures are derived from an ancestral form. The human wrist and the cow's knee are homologous, being one among many examples that could be drawn from the fact that the vertebrate skeleton has one basic pattern.

Homosexuality. A gift, a normality, a condition, a disease, a vice, a crime, a sin? Homosexuality has been all of these things at one time or another. For >sociobiologists, it has taken yet another form: it is an evolutionary advantage. Starting, as sociobiology always does, from the standpoint that all social behaviour can be understood as being controlled by genes which have evolved because they are advantageous, three quite different theories of a genetic basis for homosexuality have been put forward. First, there is the idea that there is a single gene for homosexuality. It has a curious dominance pattern, in that persons who have one such gene coupled with one gene for heterosexuality are even more heterosexually active than people with two heterosexuality genes (persons with two homosexuality genes are more or less exclusively homosexual). This means that the people with one of each gene have the most babies, and that is how the homosexuality gene continues to exist in the gene pool and to be beneficial to it (this is heterozygous advantage).

The second and third theories both relate to kin selection. A single gene for homosexuality might be of advantage in a kinship because the person with the gene is able to be a significantly useful helper of their non-homosexual siblings' reproductive performance. This can be imagined in a scenario such as a small proto-human troop, with the dominant male mating with the females and a number of homosexual males assisting the troop by keeping watch for predators, foraging and so on. The third theory is a variant of this, in that the gene is supposed to be one that induce parents to bring up some of their offspring homosexual so that they will help the non-homosexual children with child raising as above.

Any of these ideas could be true, but they are mutually exclusive, and there is absolutely no evidence for any of them. This kind of reductionist analysis of something as complex and manifestly socially conditioned as human sexual preference causes anthropologists, psychologists and others to despair.

Homosexuality, as a sexual preference for one's own sex, does not seem to occur in animals, though it is a commonplace that mock homosexual activity of various sorts does take place, e.g. heifers mounting one another when they are on heat, or male dogs mounting one another when they have nothing else to do.

Homozygote (adj. 'homozygous'; noun 'homozygosity'; from the Greek *homo*, 'same', and *zugon*, 'yoke'), an organism that, at a given locus, has two alleles the same. The term *double homozygote* refers to an individual who, at two given loci, has matched alleles. Homozygotes breed true with respect to the traits in question. Obviously any organism is a mixture, being a homozygote at many loci and a heterozygote at others, so that it does not strictly make sense to speak of a homozygous genotype if the total genotype is meant. None the less one can refer to increases or decreases in the average amount of homozygosity within populations, which is influenced by the amount of >inbreeding or >outbreeding. *See also* hemizygote.

Homunculus, the complete miniature human being that used to be thought to be contained in a sperm cell. The existence of homunculi was part of the theory of >preformation (see also pp. 4–5). The word *homunculus* comes from the Latin for 'tiny man'. Since presumably other species were believed to have their preformed miniatures too, these must have been referred to with similar names, e.g. *caniculus*, 'tiny dog'.

Horizontal transmission, the transmission of DNA by viral infection into cells (*see* transduction). Normal genetic transmission is >vertical transmission.

Hormones, any of the chemical messengers in the body. They are produced by the endocrine glands and carried round in the bloodstream to alter the behaviour of other specific target cells or tissues. Steroid hormones are small lipid molecules that enter the target cells; they include the >sex hormones. Other hormones are polypeptides (chains of amino acids shorter than proteins) which bind on to the outside of the target cell; they include >growth hormone and >insulin. Many hormones can be synthesized to provide treatment for patients with a deficiency; some can now also be produced by gene cloning. Genetic defects in a hormone system can produce either a fault in the production of the hormone or a fault in the receptor cell; both are likely to be single recessive alleles.

In insects, there are hormones that trigger the phases of metamorphosis, as well as the physiological sort described above. Plants have three types of hormones: gibberellins (stimulating stem growth), auxins (stimulating regeneration) and cytokinins (promoting cell division). *See also* allomones; pheromones.

Host. There are two uses of this word in genetics: *(a)* the recipient of a graft or transplant; and *(b)* the cell or organism that is infected by some pathogen or parasite, particularly a bacterium invaded by a virus.

Hot spot, a site in a DNA molecule where mutations or recombinations occur with much higher frequency than elsewhere.

Housekeeping genes, a gene that has a product essential for the normal metabolic requirements of any cell. These genes are active in most or all cells, in contrast to the cell-specific >luxury genes. The parts of the chromosome where they are situated show up as light-staining interbands (*see* banding).

Human evolution, or, as >Darwin had it, the 'descent of man'. The human (or *hominid*, a term that means the group including >*Homo sapiens* and its extinct human-like precursors) stock diverged from the ancestral stock of the great apes in the Miocene age, about 8 to 14 million years ago. For some reason, the environment at that time was extremely favourable to ape-like creatures, and many of them evolved and were successful all over the Asian land-mass as well as in Africa and in Europe. In about 10 million BC there was a split in the lineage, which separated the Asian apes (the orang-outan and the closely related gibbon and siamang, with their Miocene ancestor *Ramapithecus*) from the African apes, including the ancestors of humans and the African great apes. Somewhere around 8 million years ago, the African lineage also split, along ecological lines, with the forest-dwelling apes (ancestors to chimpanzees and gorillas) diverging from savannah-dwelling hominid ancestors.

The first fossils that can definitely be called hominid come from East Africa: *Australopithecus afarensis* is dated to 3–4 million BC (the best-known specimen is 'Lucy', comprising about half of the skeleton of a creature about 3ft/1m high). The genus *Australopithecus* (from the Greek for 'southern ape') stood upright, walked on its feet with no support from its hands, and was probably omnivorous (though not much of a hunter). A successful genus found widely throughout the savannah lands of Africa, it occurred in several different types (including *A. africanus*, the one ancestral to humans) whose ranges overlapped; they must have occupied different ecological niches.

However, the heavier species (*A. robustus*) eventually lost out to the lighter type represented by the lineage that led from *Homo habilis* (the first fossil hominid to show substantial enlargement of the brain, and to use tools) to *Homo erectus*. This last was a very successful species: starting from East Africa in about 1.6 million BC, within 600,000 years it had reached South-east Asia, within another 500,000 years, it had travelled to China (where the remains of the famous Peking Man were found) and perhaps to Europe. *Homo erectus* had a brain at least two-thirds the size of that of modern humans, and had a skeleton almost entirely the same as the modern version (evolutionary improvements having been made particularly in the legs and pelvis); it was able to use fire and a variety of tools, and was probably an efficient hunter.

The era between 400,000 and 200,000 years ago, during which >*Homo sapiens* emerged, was characterized by the development of large regional differences. One of the predominant types was *Home sapiens neanderthalensis*, Neanderthal Man, of which the type specimen came from Germany but which was found throughout Europe, the Middle East and North Africa; similar early *Homo sapiens* subspecies (e.g. *H. s. rhodesiensis*) were found elsewhere. (It is not true that Neanderthal Man walked with bent knees; this misunderstunding arose because one of the specimens had severe arthritis.)

Truly modern *Homo sapiens* first appeared about 120,000 years ago, with the earliest specimens coming from Ethiopia and South Africa. Within 80,000–90,000 years, *Homo sapiens* was established in Australia; the colonization of the Americas by people from Asia crossing the land-bridge at the Bering Straits probably occurred at about the same time, and the whole of the two American continents were populated by 10,000 BC. The earliest fully modern fossil found in Europe is 'Cro-Magnon Man', about 25,000 years old.

Human Genome Project. *See* pp. 36–8.

Hunter's syndrome, a human disease caused by an X-linked recessive allele. The biochemical problem (as with >Hurler's syndrome) is that the patient is unable to break down mucopolysaccharides, large compounds of sugars that combine with proteins and are important in nerve cell function. A build-up of these molecules leads to mental deterioration and deafness, as well as various disfiguring symptoms. Women in families where the disease has occurred can be tested to see whether they are carriers; and prenatal diagnosis is reliable.

Huntington's chorea, a human disease caused by a single dominant allele. The underlying mechanism is not known, but the disease is a distressing one consisting of involuntary movements and mental deterioration; it does not appear until the sufferer is 40 years old or more, and is at present incurable, leading to death within five to ten years. Before diagnosis, the symptoms may be mistaken for drunkenness, and this can lead to much distress for sufferers and their families, especially if they lose their jobs because of it. Diagnosis is no comfort, however, for not only is it the death sentence on the person who has the disease, it may mean the same for the children that they had before they knew about it.

It is now possible to use a genetic marker on chromosome 4 to detect with about 95 per cent accuracy whether a person whose parent had Huntington's chorea is going to develop the disease. However, this can only be done in families with enough members in preceding generations available to provide DNA samples, and only if the relative with the disease has a marker gene that is different when compared to those of the ones without Huntington's chorea. Similarly, a prenatal test can detect whether a foetus is risk-free or not. The decision whether or not to abort a high-risk foetus is complicated by two things: *(1)* even if it has the gene, it will have at least 40 years of normal healthy life; and *(2)* a cure for the disease could be found during its lifetime. In white populations, about 1 in 10,000 has the gene for Huntington's chorea.

Hurler's syndrome, a human disease caused by a recessive allele. The biochemical problem is the same as in >Hunter's syndrome, but the effects are worse, with dwarfism and heart problems leading to death at the age of 8–10. Carrier females can be identified biochemically, as can affected foetuses. About 1 in 100,000 babies has Hurler's syndrome, with little difference in frequency in different ethnic populations.

Huxley, Thomas Henry (1825–95), British biologist. Huxley trained as a doctor and served as a naval surgeon. He went on the voyage of HMS *Rattlesnake* round Australia (1846–50), and did research on plankton, which established his scientific reputation (he won the Royal Society's Gold Medal for it). When first he read *On The Origin of Species*, he exclaimed, 'How silly of me not to have thought of that!' and he became a passionate supporter of >Darwin's theory of evolution, earning himself the nickname of 'Darwin's bulldog' (*see* p. 18). In *Man's Place in Nature* (1863), he made the first explicit statement of the non-unique biological position of humans. Huxley refused to believe anything not susceptible to logical analysis, and therefore rejected both atheism and belief in God. He became a member of the London School Board and was highly influential in getting rid of rote learning and substituting basic skills of numeracy. His son Leonard and grandsons Julian, Aldous and Andrew have all been intellectually distinguished.

Hybrid, broadly, the offspring of any cross between two (fairly pure) strains. In a narrower sense, the term is used to mean the offspring of a cross between parental strains that are homozygous at the given locus or loci, such that the hybrid individual is heterozygous for these (*see* F_1; monohybrid; dihybrid; trihybrid). A hybrid can also be a cross between different species or genera, e.g. a mule or hinny.

Hybrid DNA, DNA that is artificially made *in vitro* by mixing two strands of DNA (or one of DNA and one of RNA) of different genetic origins (*see* heteroduplex). It is not the same as >recombinant DNA.

Hybrid vigour, the phenomenon (technically called 'heterosis') in which the hybrid is more vigorous (larger, more prolific, more disease-resistant, etc.) than either of its parental strains. The genetic explanation is that each of the parental strains has become inbred to the point where, at a significant number of loci, recessive alleles with a bad effect on the genotype have become fixed (i.e. all the individuals are homozygous at these loci). The hybrid between two such strains has a good chance of having dominant alleles from the other strain to cover the poor effects of the recessive in at least some of the loci concerned. *See* F_1.

Hybridoma, an artificially produced hybrid cell, combining a normal lymphocyte and a myeloma (cancerous) lymph cell, and used in the preparation of monoclonal antibodies.

Hypercholesterolaemia, familial, an inherited condition in which too much cholesterol is deposited in the blood vessels, leading to a high risk of coronary artery disease. The physiological problem is that the affected person has too much of a type of protein called LDL which is involved in cholesterol deposition; the genetic factor is a defective gene for the receptor molecule that normally removes excess LDL and therefore also excess cholesterol. This defective gene is inherited as a dominant (there may in fact be a number of different mutations of the receptor gene, all of which have the same effect). Hypercholesterolaemia occurs in white populations at a frequency of about 1 per 500, and is responsible for about 5 per cent of early deaths from heart attacks. People who have the gene will not inevitably get heart disease, as a low-fat diet has a strong preventive effect, particularly if started early in life.

Hypersensitivity, over-reaction by the >immune system. It involves the mounting of a large-scale immunological defence against antigens that are in low concentration and are possibly not harmful to the organism in that low concentration though they would be at a higher one. *See* allergy.

Hypha (pl. hyphae), a filament that forms the substance, or mycelium, of some fungi. Hyphae cells are haploid and can produce asexual spores called >conidia. *See also Neurospora crassa.*

I

I_1, I_2, I_3, etc., the first, second, third, etc. generations obtained from an outcross, i.e. a cross between strains that are themselves already heterozygous. Compare this to >F_1 where the initial cross is between inbred strains.

Identical twins. *See* monozygotic twins.

Idiogram, a formal representation of all the chromosomes in an individual's >karyotype, usually consisting of photographed chromosomes from a single cell-division, which have been cut out and arranged in order. Preparation of an idiogram is an essential step in diagnosing >chromosome abnormalities.

IgA, IgD, IgE, IgG, IgM, the various different types of >immunoglobulin found in the human immune system.

Immigration. *See* migration.

Immune system, the system by which the body overcomes infections and other invasions by foreign bodies. *Natural*, or *non-specific*, *immunity* works by phagocytosis (literally, 'cell-eating', in which cells surround and dissolve particles or bacteria) and by the action of the protein interferon (the mode of which is still not clear). *Specific immunity*, i.e. a response to one particular infective agent (all of which, whether viruses, bacteria, parasites, or non-living particles are known as antigens in this context), is most highly developed in vertebrates, including humans. It is of two kinds: *humoral immunity*, in which antibody molecules circulate in the lymph and blood; and *cell-mediated immunity*, in which lymph cells directly bind to the antigens.

In humoral immunity, the arrival of the antigen stimulates the appropriate line of >B lymphocytes in the bone marrow (*see* clonal selection theory); these cells, and their descendants in the lymph nodes and spleen, produce the specific antibody molecules which bind on to the antigens, either making them incapable of infective action or holding them in lumps so that they can be easily engulfed and dissolved by the large macrophage cells (this is facilitated by a group of proteins collectively known as >complement, activated by the antibody–antigen complex). Cell-mediated immunity is similar, with the antigen binding being done by >T lymphocytes which originate in the thymus gland. The humoral immunity produced by circulating antibody is most effective against infections by bacteria and viruses while they are outside the cells; cell-mediated immunity is better against viruses inside cells, parasites, cancer cells and foreign tissue. *See* auto-immune diseases; immunization.

Immunization, the technique of artificially creating >active immunity by injecting the patient with non-virulent antigens that resemble the disease-causing pathogens (*see* vaccination). Some diseases have a more lasting immunity than others: the acquired immunity to measles and mumps is generally lifelong, but in some cases (e.g. flu) it is very short-lived. This is probably because of changes in the virus responsible, which affect its antigenic properties.

The term is also used for giving >passive immunity by the injection of antiserum. Immunization can also mean any experimental process in which animals are injected with some substance to which they acquire immunity.

Immunoglobulins (abbreviation Ig), the proteins that are produced as antibodies in the immune system. There are several types, each consisting of two pairs of chains, two heavy (H) and two light (L). Each Ig chain, whether heavy or light, consists of a length that is constant (the *C region*, occupying about half of an L and three-quarters of an H chain), a joining region (*J region*), and at the other end a variable part (*V region*). Specificity to the antigen is due to the sequence in the variable regions, and to the combinations of the variable-region sequences in the heavy and light chains. Although each immunoglobulin-producing cell has the same DNA sequence for immunoglobulin, each will have a different phenotype, in that it suppresses all but a few sequences to produce its own unique combination of V and J regions. The five types of immunoglobulins found in humans are:

- *IgA*, found in various fluids that protect mucous surfaces (in tears, gut and urinogenital fluids and sweat).
- *IgD*, found on the surface of B lymphocytes, especially in the foetus.
- *IgE*, effective against helminth (worm) infections and involved in the >hypersensitivity reaction (*see* allergy).

- *IgG*, the main active antibody, which can also be passed through the placenta so that the foetus can receive some immunity from its mother.
- *IgM*, the largest (consisting of five of the normal immunoglobulin units) and the first to be made after an antigenic challenge.

Immunosuppression, artificial suppression of the immune system, used to help to prevent rejection of a transplant. Drugs such as cortisone decrease the activity of the T-lymphocytes. There is always the risk that a patient who is receiving immunosuppressive drugs to protect a transplant will die from an infection; in the early days of heart transplants there was a high rate of deaths from pneumonia because of this.

in vitro, from the Latin, 'in glass'. A term for processes that happen in a controlled environment outside of a living organism, usually in tissue culture. It is strictly speaking an adverb but is often used adjectivally. The opposite is >*in vivo*.

***in vitro* fertilization**, the technique of fertilizing one or several ova with sperm outside the body of the female, in a glass culture dish (hence the term *in vitro*, 'in glass'). It is used as a treatment of infertility, where the problem is blocked or absent Fallopian tubes (which lead from the ovaries to the uterus) which make it impossible for an egg to pass from the ovary to the uterus (or for sperm to travel from the uterus to meet the egg in one of the tubes). The first baby to be conceived by *in vitro* fertilization was Louise Brown, born in 1978. She was the first of thousands of normal babies who prove that there is not, as was at first feared, a risk that *in vitro* babies could be abnormal.

To increase the chances of pregnancy, it is usual to give the mother a >fertility drug to increase ovulation so that several eggs can be collected, fertilized and implanted at once; this has resulted in many cases of mothers becoming pregnant with more babies than were intended (*see* multiple births). The eggs must be removed surgically from the mother and put into a fluid in a culture dish. Shortly before the eggs are collected, the father has to provide a fresh sample of sperm by masturbating. Fertilization takes place in the fluid, and progress of the embryo is carefully watched under a microscope, until three or four cell divisions have taken place (within 48–72 hours); then the embryo is transferred into the mother's uterus. The most uncertain part of the procedure is whether the embryo will successfully implant itself in the uterus, which only happens in one out of five or six attempts. *In vitro* fertilization is also used when the mother is unable to ovulate, and an egg from another woman is used.

The technique has had the support of public opinion since it was introduced, except that there is much misgiving about what should be done with embryos that are created *in vitro* but are surplus to the number wanted for implantation; they can be frozen and used if a second attempt at implantation has to be made, but if this does not happen, is it acceptable either to throw them away or to use them for experimentation? In Britain the Human Fertilization and Embryo Research Act 1990 has legalized experimentation on embryos up to 14 days after fertilization.

in vivo, from the Latin, 'in a living thing'. A term used for processes that happen in a living organism. It does not imply that the process described is as it occurs in natural conditions, because the word is applied to what happens in experiments; it is used solely in contrast to what happens in tissue culture (*see in vitro*). It is strictly speaking an adverb, but is often used adjectivally.

Inborn errors of metabolism, a group of disorders in which the basic cause is the lack of an enzyme involved in normal metabolism. These disorders are almost always inherited as recessives. For example, if the normal gene coding for the enzyme is T, and the mutant gene is t, a heterozygous person Tt will be healthy because the T gene can produce sufficient quantities of the enzyme; whereas the person with genotype tt will be without the genetic means to produce the enzyme and will therefore have a block in a metabolic pathway. The term 'inborn error of metabolism' was coined by Sir Archibald >Garrod who identified four of them in 1903; over 100 such conditions have now been described. Most patients with these disorders can be treated with diets that avoid taking in the substance that cannot be metabolized. *See* alkaptonuria; cystinuria; phenylketonuria.

Inbreeding, mating between related individuals. It is a system that increases the amount of homozygosity within the genotypes of the succeeding generations. Under the most extreme form of inbreeding – self-fertilization – the proportion of heterozygotes is halved at every generation. In the case of the next-closest form of inbreeding – parent–child or brother–sister – the reduction is only half as fast. Inbreeding has a bad reputation, partly arising from its association with >incest (which is a social concept, not a biological one). In fact, whether or not inbreeding produces bad results is entirely a matter of how many deleterious recessives there are in the stock to begin with: a stock with a large number will decline as these come into the homozygous state in more and more individuals, but if there are few or none, there will be no such effect. Indeed, if the recessives in the original stock produce beneficial effects, an inbred strain will actually improve, a phenomenon that has been seen again and again with laboratory animals (e.g. the standard strains of laboratory mice are all much bigger and more prolific than wild mice). It is also a principle much used in agricultural breeding (where it is known by the more positive name of >line breeding).

Inbreeding, coefficient of; a measure of the increase of >homozygosity as a result of inbreeding, as compared to what would be expected with random mating. The coefficient of inbreeding, F, devised by Sewall Wright, can be taken to be either a property of inbred populations, being the average proportion of homozygosity in that population, or a property of inbred individuals, being the proportion of loci at which the individual carries two alleles identical by descent.

Incest, sexual relations between closely related individuals. Almost all human societies forbid incest, although the relationships prohibited vary: parent–child and brother–sister incest is practically always forbidden, but sexual relations between uncle and niece, aunt and nephew or cousin and cousin are viewed very differently, being sometimes banned, sometimes encouraged, sometimes obligatory. The genetic consequence of incest is to make it more likely that any deleterious recessives in the family will come into the homozygous state (*see* inbreeding). The sociobiological view of incest is that since the genetic consequences of close inbreeding are unfavourable, any gene that caused aversion to mating with one's close relatives would be favoured by natural selection. The existence of such a gene is, of course, unproven, notwithstanding the fact that our nearest evolutionary relatives, the great apes, sometimes (but not always) avoid incest. The sociobiological theory does not explain why there should be so many societies in which first-cousin marriage, which is quite close inbreeding and known to cause an excess of genetic disorders in the offspring, should be encouraged. The anthropological view of incest, given the diverse forms in which it occurs, is that it is a cultural not a biological phenomenon. (And that it is absurd to worry that children born by artificial insemination from the same donor may unwittingly marry and so commit incest.) It is certainly the case that a great deal of illicit incestuous mating goes on in developed societies, mostly father–daughter or father–stepdaughter. The fact that the latter comes under the incest taboo fits with either the sociobiological or the anthropological viewpoint, but the fact that there is so much of this non-admitted activity is strongly supportive of the idea that the incest taboo is under the control of social sanctions rather than genes.

Inclusive fitness, the total >fitness (i.e. comparative reproductive performance) of an individual, including its own reproduction and that of its near relatives (their contribution diminishing with the distance of the relationship). It is an important idea in the arguments surrounding >kin selection.

Independent assortment, the principle that genes are inherited independently of one another; >Mendel's Second Law. When Mendel was doing his experiments on peas, he found that the inheritance of one trait was independent of another: a pea plant might have purple or white flowers, but the colour of the flowers had no effect on whether the plant produced round or wrinkled peas. The underlying mechanism is that genes will behave in this way if they are located on separate chromosomes. The diagram below shows a very simplified example, in which there are two pairs of chromosomes.

If the gamete on the far right of the F_1 row is fertilized by a plant from

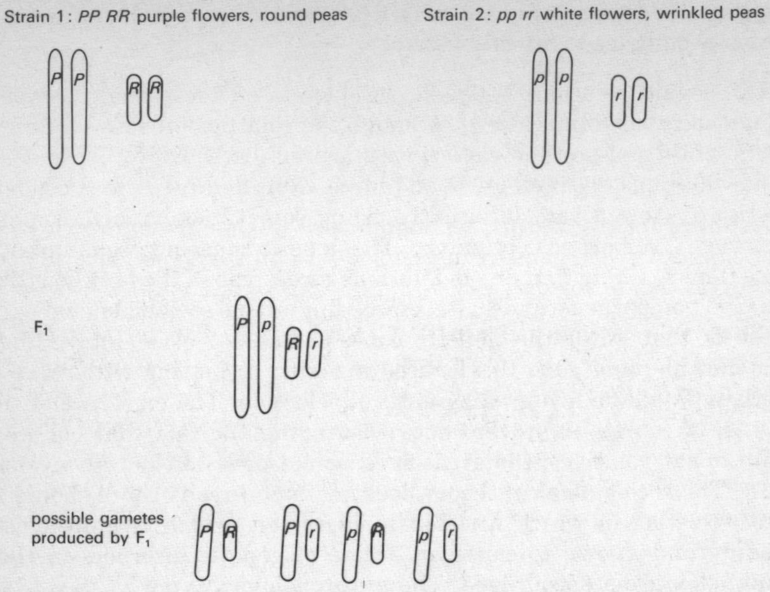

Strain 1: *PP RR* purple flowers, round peas

Strain 2: *pp rr* white flowers, wrinkled peas

F$_1$

possible gametes produced by F$_1$

Strain 2, the result will be a plant with genotype *pp Rr*, with white flowers but round peas, showing that the two traits are inherited from the original strains independently of one another. If, however, the genes are on the same chromosome, they are said to be in >linkage, and they do not show independent assortment. Mendel was very lucky that the loci of the seven traits he was investigating were on the seven different pairs of chromosomes; the odds of this happening are 166 to 1 against.

Inducer, a molecule that binds to a repressor to turn on the transcription of an inducible >operon. The inducer is usually a small molecule, smaller than the repressor (a protein), but able to alter the latter's configuration so that it dissociates from the operator region and allows transcription to proceed. The inducer is often the substrate of the enzyme whose production is being induced. *See* gene regulation.

Inducible enzyme, an enzyme which is produced only when its substrate is present to act as inducer. The opposite is a repressible enzyme. *See* operon; gene regulation.

Infertility, inability to produce viable offspring. It has been estimated that about 10 per cent of human sexual partnerships are infertile. The problem may be in either partner or (more rarely) may be a genetic incompatibility between them. The commonest causes are (in men) low sperm count or non-viable sperm, and (in women) blockage of the Fallopian tubes. Treatments are available for all forms of infertility. Sperm problems can be treated by artificial insemination; Fallopian tube blockage by *in vitro* fertilization; failure to ovulate by egg donation; failure to sustain pregnancy by surrogacy. *See also* sterility.

Insertion, the addition of one or more base-pairs into a DNA sequence. The effect of a small insertion may be a >frameshift mutation, but a larger insertion can upset gene regulation around it, either by activating genes that should be inactive or vice versa. *See* insertional inactivation.

Insertional inactivation, a technique used in genetic engineering to prevent expression of a gene by inserting a foreign DNA sequence adjacent to it or in its coding region.

Instinct, a behaviour pattern that is inherited. The term is not much employed by ethologists and psychologists, as it has been used in so many ways that its meaning has became elusive. But there is still a useful distinction to be made between *instinctive* behaviour, where the stereotyped pattern of actions is genetically programmed and is prompted by a specific stimulus and *learned* behaviour which, however typical of the species, is the property of the individual who has learned it. This is not to deny that some instinctive behaviour patterns require a certain learning process; in many singing birds, the urge to sing is innate but the bird needs to hear the song of its particular species. The mode of inheritance of instinctive behaviour patterns is still unknown. The pioneers of the study of instinct were Konrad >Lorenz and the Dutch ethologist Nikolaas Tinbergen.

Insulin, the hormone involved in sugar (more specifically glucose) metabolism. Insulin deficiency causes >diabetes, which is treated by diet or with injections of insulin. Formerly, insulin had to be extracted from pigs or cattle, but human insulin can now be produced by gene cloning.

Intercross, a cross between two individuals or strains that are themselves hybrid. *See* F_2.

Interferons, a group of proteins that are produced by cells in response to a viral attack. They seem to enable other cells to resist viruses, but the mechanism is not understood. It is believed that interferons may be anticancer agents, if they are able to inhibit oncogenes in the same way as they do viruses, but clinical trials have not confirmed this. Genes for human interferons have been cloned so the proteins can now be produced in bulk.

Interphase, the stage on the >cell cycle between divisions. This is the phase during which the chromosomes are active in synthesizing proteins.

Intersex, a broad term covering various conditions in which an individual does not have normal reproductive organs but is not a hermaphrodite. Male pseudohermaphroditism is the condition in which there are testes (probably internal) but the external genitalia appear more or less female; the cause can be chromosome abnormality, a single gene (as in testicular feminization) or hormone imbalance *in utero* (*see also* Klinefelter's syndrome). Female pseudohermaphroditism, in which there is a combination of normal ovaries and more or less male external genitalia, is not caused by chromosome disorders, but by a series of recessive alleles, with a combined frequency of about one in 7000 births. Hormone imbalance *in utero* can also be responsible.

Interspecific competition, competition between different species for a resource. It may lead to the extinction of one of them, or at least to its exclusion from a particular area.

Intrachromosomal recombination, exchange of sequences of DNA between sister chromatids (i.e. two halves of the same chromosome) during meiosis. Because the sister chromatids are identical, there is no genetic effect, unlike in the normal process of recombination, when non-sister chromatids from homologous chromosomes exchange genetic material.

Intraspecific competition, competition for a resource between members of the same species. This is the 'struggle for existence' that Darwin and Wallace identified as the driving force of evolution.

Intron, a sequence of DNA that does not code for a gene product (in contrast to an >exon). The sequence is initially transcribed into RNA, but all the introns are cut out before the messenger RNA is finalized. This is called 'intron splicing'. Introns do not occur in prokaryotes, but they are very common in eukaryotes, though their function is obscure.

Inversion, a form of mutation in which a sequence of a chromosome is removed and replaced the other way round. Because the genes remain the same, there is usually no genetic consequence, but sometimes a gene's activity may be altered due to its being separated from its normal neighbours, which are be part of its control mechanism: *see* position effect. Inversions can be either *pericentric*, i.e. involving the >centromere, or *paracentric*, involving only one of the arms.

Invertebrates, all animals without backbones. These include sponges, jellyfish, sea anemones, worms, molluscs, starfish, spiders, crustaceans and insects. *See* classification.

IQ (intelligence quotient), a measure of mental capacity, defined as a person's mental age × 100 divided by its chronological age. Thus a child of 10 with a mental age of 9½ has an IQ of 95. The idea of intelligence as being something that could be measured and the result represented by a single figure – even one that, like IQ, is derived from a whole series of tests – has been criticized from the outset. The contents of the tests used by pioneers of IQ testing are to modern minds quite ludicrously socio-culturally biased. None the less the fact that IQ was a 'measurement' gave it a spurious reality and respectability, and battle-lines were drawn up as to whether the observed differences between children (and groups of children) were genetically determined or not.

The important thing is not to try to discuss whether 'nature' is more important than 'nurture' – the geneticist's position is that this is a non-question (*see* nature versus nurture; heritability) – but to ask why the matter has been so hotly and emotionally debated. Broadly, hereditarians (i.e. those who subscribe to the 'nature' side of the argument) have been politically of the right, and have been anxious to prove that their monopoly of educational and employment opportunities has been justified by their superior genetic endowment. Environmentalists (adherents of 'nurture') have been of the left, viewing the human infant as a *tabula rasa*, and keen to show that inequalities of intelligence are the result rather than the cause of inequalities of opportunity. Legitimate methods of research have included the study of adopted children (who should show more resemblance to their natural or adoptive parents, according to hereditarians and environmentalists respectively), and the study of identical twins, measured both against non-identical twins, and, if reared apart, against the different environmental conditions (*see* twin studies). The many studies along these lines have, not surprisingly, come up with different conclusions. Sir Cyril >Burt was so anxious to prove his hereditarian position that he falsified his research, making up pairs of twins that did not exist. In the late 1960s there was an attempt by the American psychologist Arthur Jensen and others to show that there were differences in average IQ between races, specifically between white and black children in the United States. It was bad science, and bad politics, but what was perhaps saddest of all was the response by people who should have known better: rational scientists and others apparently felt that all that was needed to meet this idea was a torrent of abuse, denouncing the perpetrators as fascists.

Isochromosome, an abnormal chromosome in which two identical arms have become joined at the >centromere. Thus half of the genes that ought to be on the chromosome are present in double dose, while half are missing altogether.

Isoenzymes (also isozymes), a set of enzymes all carrying out the same chemical reaction. Several isoenzymes may be present within the same organism, as they are coded for by genes at different loci. Compare >allozyme.

Isogamy, the production of two equal-sized types of gametes by a species, i.e. the male and female gametes are identical. This is a primitive feature, not found in the higher eukaryotes whose sperm and ova are specialized in shape and function (*see* anisogamy).

Isolation, in genetic terms, the situation in which a given population is unable to exchange genes with any others, owing to barriers (geographical or other). *See* isolating mechanism.

Isolating mechanism, any barrier that prevents interbreeding between one population and another. The simplest form is a geographical barrier, but a behavioural, physiological or anatomical one is just as effective in separating populations, which is the first step towards the formation of new species.

J

J region, or joining region, the part of an >immunoglobulin molecule that lies between the constant and the variable regions.

Jacob, François (1920–), French microbiologist; director of the Pasteur Institute since 1982; co-proposer of the >operon mechanism of gene control. In 1961, Jacob and E. Wallach announced the discovery of extranuclear genetic elements that they called episomes (now known as plasmids), and in the same year, Jacob and Jacques >Monod proposed the theory that gene action is controlled in a unit called an operon, and also predicted the existence of the type of RNA molecule now known as messenger RNA. For this work Jacob shared the 1965 Nobel prize for physiology or medicine with Monod and André Lwoff.

Johannsen, Wilhelm Ludwig (1857–1927), Danish geneticist; inventor of the word 'gene'. Too poor to go to university, Johannsen became a pharmacist, but through his private researches, he became the leading theoretical geneticist of his time. He began to study heredity in the 1890s, and was influenced by Galton; he proposed a particulate theory of inheritance which he published before the rediscovery of Mendel's very similar theory. He was the first to make the distinction between the genotype and the phenotype; in the same paper, published in 1909, he suggested the word 'gene'. Johannsen later became rector of the University of Copenhagen.

Jumping genes, genes that move within chromosomes. They influence the regulation of gene activity, probably by physically removing a gene from its promoter (*see* operon). Jumping genes were discovered by Barbara >McClintock. *See also* transposon.

Junk DNA (also known as >selfish DNA), sequences of DNA in the genome that have no apparent genetic function.

K

***K* selection**, the type of natural selection that operates when the environment is relatively constant and the population is at or near the carrying capacity. In these circumstances, high reproductive rates will not pay off, but good adaptation, particularly efficency in food and other resources, will be favoured. Longer parental care and small numbers of offspring, a high proportion of which are successfully raised are also favoured. *See also* *r* selection; MacArthur, Robert Helmer.

Karyotype, the chromosomal constitution of an individual as seen in the nucleus of a somatic cell. The chromosomes are classified by their size and centromere position, and for convenience are usually shown in a specific order in an >idiogram (this is also sometimes referred to as the karyotype). The karyotype for a normal human male would be notated '46, XY'. (Note the ambiguity: this means 46 chromosomes in total – i.e. 44 autosomes and two sex chromosomes – but it looks as if it could mean 46 autosomes plus two sex chromosomes.)

Khorana, Har Gobind (1920–), Indian-born American molecular biologist; professor of biology and chemistry at the Massachusetts Institute of Technology since 1970. By 1961 Francis >Crick and Sydney >Brenner had found that the >genetic code consists of cordons, each of which were three bases long. Khorana applied his skill in synthesizing biological molecules to the problem of deciphering the code: by creating each codon artificially, he was able to show which one coded for which amino acid. He then achieved the first-ever synthesis of a gene, when he reproduced artificially the DNA sequence of an >*Escherichia coli* gene. Khorana shared the 1968 Nobel prize for physiology or medicine with Marshall >Nirenberg and Robert W. Holley.

Kilobase, literally, 1000 bases. A unit of length of DNA or RNA, used in describing genes or shorter sequences. When measuring double-stranded DNA, it is actually base-pairs that are counted, so that the unit is comparable with counting bases in a single-stranded molecule.

Kin selection, natural selection that acts upon a group of kin rather than upon the individual, and produces evolutionary changes of the sort that favour kinship groups rather than individuals. After the demise of the theory of >group selection, kin selection was proposed in 1964 by W. D. Hamilton as the best explanation of how altruism could have evolved. It is true that 'altruistic' genes will be preferentially represented in the succeeding generation as a result of altruism towards close kin, given that relatives share many genes. Full siblings have 50 per cent of their genes in common, so that if the benefit to sister B for a given amount of effort is more than double the benefit to sister A, it pays sister A to help sister B. In evolutionary terms, the help that counts is help towards reproducing successfully, i.e. towards getting one's own (or one's sister's) genes preferentially represented in the next generation. By the same calculation, it pays to help a first cousin (one-eighth of genes in common) if the benefit to her is more than eight times the benefit of the same effort to oneself. From this, it appears that kin selection is likely to be much more effective between very close kin, and sociobiologists point out that small groups of close kin are the actual breeding units of very many species.

Klinefelter's syndrome, the condition in which a male person has genotype XXY (i.e. two X chromosomes and one Y chromosome). Boys with Klinefelter's syndrome look normal until puberty, but in adulthood, they have some female features such as enlarged breasts and sparse body hair, and their testes are small and never produce sperm. These features are due to the conflicting genetic information in the sex chromosomes as to which sex the individual is. Less obviously, they are taller than average, and a significant percentage are mentally retarded, for which there is no physiological explanation. The extra chromosome usually comes from the mother (i.e. she produced an ovum with two X chromosomes), and there is an increasing risk of this with maternal age, but in a few cases, the extra X comes from the father (i.e. he produced a sperm with an X and a Y). Klinefelter's syndrome is the commonest of the sex chromosome disorders, occurring at a rate of about 1 per 1000 male babies. Hormone replacement therapy, consisting of injections of the male sex hormone testosterone, is effective, particularly if it can be started at puberty (though many cases are not noticed during childhood).

Kornberg, Arthur (1918–), American biochemist; professor of biochemistry at Stanford University, California since 1959. Kornberg specialized in enzyme biochemistry, and isolated the enzyme DNA polymerase I from >*Escherichia coli*. This is one of the two enzymes that normally build the second strand of DNA during replication, but Kornberg found that it could also assemble a DNA molecule >*in vitro*, using only unconnected nucleotide molecules, provided that there was some natural DNA in the mixture to act as a template. This became an important technique in later DNA research. Kornberg shared the 1959 Nobel prize for physiology or medicine with Severo Ochoa.

L

Labelling, a technique for tagging molecules by introducing a radioactive isotope (e.g. carbon-14, iodine-125) into their structure. The presence of the radioisotope does not affect the chemical activity of the molecule. The fate of the labelled molecule(s) can then be tracked, in either in >*in vivo* or >*in vitro* experiments, by autoradiography or scintillation counting.

***Lac* operon**, the group of genes in *Escherichia coli* that controls the metabolism of lactose. It was the first >operon to be discovered, and was found by Jacob and Monod in 1961. They identified the following parts. First is the gene *lacI*, which codes for the >repressor; then the control region, consisting of the >promoter region *lacP* and the operator region *lacO*; lastly come the three structural genes *lacZ*, *lacY* and *lacA*, which code for the three enzymes needed for metabolizing lactose.

In the absence of lactose (i.e. in normal conditions for *E. coli*), a molecule of the repressor protein is bound to the operator region, preventing RNA polymerase from proceeding downstream from the promoter where it attaches. When molecules of lactose attach to the repressor, the repressor is lifted off the operator and transcription of the three enzymes occurs. One of these (Y) transports lactose through the cell membrane; another (Z) hydrolyses it; the function of the third (A) is not clear. When all the lactose in the cell has been metabolized, the repressor binds once more to the operator and synthesis of the enzymes stops. The system of having the substrate of the enzymes act as the inducer of the operon is commonly found. Jacob and Monod put together this scheme of gene regulation by deduction from the effects of various mutations that interfered with its normal working.

Lamarck, Jean-Baptiste Pierre Antoine de Monet, Chevalier de (1744–1829), French biologist, one of the first people to propose a theory of evolution. After a medical training, Lamarck became interested in botany and wrote *Flore française*, proposing a system of classifying plants by a succession of either/or divisions. This impressed the influential biologist Comte >Buffon who, in 1781, helped Lamarck to get a job as botanist to the king. After the Revolution, in 1793, Lamarck became professor of zoology at Paris, where he did important work in the >classification of the animal kingdom. He was the first person to notice that some animals have backbones while others do not, and he invented the term 'Invertebrata' for the latter. He also identified and named the Crustacea (crabs, lobsters, woodlice, etc.), Arachnida (spiders, scorpions, mites, etc.) and Annelida (earthworms, lugworms, leeches, etc.). He is remembered for his ideas on evolution. He was quite correct in following Buffon and Erasmus Darwin in saying that species are not fixed but have evolved from earlier forms (he based this on the fact that domesticated plants and animals are changed from the wild varieties). But his most important misconception, and the one that bears his name as 'Lamarckism', was that acquired characteristics can be inherited (*see* pp. 9–10 and 14, and Lysenko, Trofim Denisovich).

Lambda, one of the most intensively studied of the >bacteriophages. It attacks >*Escherichia coli*. The genome is double-stranded DNA, but with single-stranded, mutually complementary 'tails' (*see* sticky end) that join up to make a circular DNA after entry into the host cell.

Lampbrush chromosome, an unusual type of chromosome found in the oocytes (immature eggs) of newts and other amphibians. There are two chromatids which form many paired loops joined back to the chromosome core. The loops are segments of DNA that have become decondensed so as to be transcribed, and each one probably represents a single gene.

Landsteiner, Karl (1868–1943), Austrian-born American immunologist. Between 1900 and 1902 Landsteiner discovered the >ABO blood group system in humans. Later, in 1927, he discovered the similar >MN blood group system, which is not as important in practical terms because blood transfusions are not affected by it. For these discoveries, he was awarded the 1930 Nobel prize for physiology or medicine. In 1940, at the end of his long career, he and his colleague A. S. Wiener discovered the >Rhesus blood groups.

Late genes, genes that are expressed late in the process of a virus infecting a bacterial cell. They usually code for the protein(s) that make up the virus's coat. *See also* early gene.

Laurence–Moon–Biedl syndrome. *See* obesity.

LD50, a term used to denote the dose of a substance that will kill 50 per cent of the organisms receiving it. LD50 tests are used on laboratory animals for testing the toxicity of drugs, toilet preparations, cosmetics, etc. designed for humans. When a group of bacteria or larger organisms are used, which are well matched for size (so that the relative dose they receive is comparable), the survivors may well be genetically different from the non-survivors, so that the result says as much about genetic resistance as it does about the substance being tested.

Leader sequence, the part of a molecule of messenger RNA between the 5′ end and the start of the coding sequence. It contains sequences that are concerned with binding to the ribosome.

Lederberg, Joshua (1925–), American geneticist; professor of genetics at Stanford University, California, 1959–78, then president of Rockefeller University, New York. Working with E. L. >Tatum, Lederberg discovered that in some strains of >*Escherichia coli* sexual conjugation takes place (up till then it had been thought that bacteria could only reproduce by simple division). This made it possible to map bacterial chromosomes, and was the starting point for bacterial genetics. Then, with N. D. Zinder, he discovered transduction, the process by which a virus can transfer a piece of genetic material from one bacterium to another. He also found that a B lymphocyte produces only one antibody, which was the basis for the development of monoclonal antibodies. Lederberg shared the 1958 Nobel prize for physiology or medicine with Tatum and George >Beadle.

Leeuwenhoek, Antony van (1632–1723), Dutch naturalist; pioneer of the microscope. Van Leeuwenhoek had no formal training in science. Compound microscopes had already been invented by his time, but he used instead single lenses with a very short focus (i.e. magnifying glasses) with which he obtained magnifications as great as × 250. (It is believed that he had the advantage of unusually good close-range vision.) He gave the first descriptions of red blood corpuscles, >sperm (which he supposed contained homunculi – *see* p. 5), bacteria, protozoa and rotifers. All these, and many other observations, were described in his 375 letters to the Royal Society in London (of which he was elected a Fellow in 1680). Perhaps his most important contribution was his study of the life cycle of the flea, which he announced to be ‘endowed with as great perfection in its kind as any large animal’. His work proved that fleas reproduce sexually like any other insect or animal, and were not spontaneously generated, as was believed at the time (*see* p. 7).

Left-handedness. *See* handedness.

Leptotene, one of the stages in >meiosis (one of the two processes of cell division). *See* diagram at meiosis.

Lesbianism. *See* homosexuality.

Lesch–Nyhan syndrome, a human disease inherited as an X-linked recessive. It therefore affects boys who have inherited it from carrier mothers. The actual mutation is a failure to produce the enzyme hypoxanthine phosphoribosyl transferase (HPRT), which is part of the process for metabolizing purines; build-up of only partly metabolized purines leads to mental retardation and cerebral palsy. There is also a bizarre and totally unexplained behavioural effect: boys who suffer from this syndrome have a compulsive urge to bite their own lips and fingers, and if they are not restrained, they will actually bite pieces off. The condition is very rare. Female relatives of a sufferer who may be carriers can have a foetus prenatally diagnosed, as the lack of HPRT can be detected from conception. Lesch–Nyhan syndrome is a candidate for gene therapy, as the normal gene for human HPRT has been isolated and cloned.

Lethal mutation, any mutation that has such a severe effect on the organism as to cause early death. From the genetic point of view, a mutation is considered 'lethal' if the death occurs at any time before the individual reaches reproductive age. Common usage in relation to humans, however, includes only those mutations that cause death before birth or in early infancy. Lethal mutations can be dominant, recessive or X-linked, and there are some known as >semi-lethals. Lethal mutations are quite common in humans: about 20 per cent of pregnancies end in spontaneous abortion (miscarriage), with either lethal mutations or chromosome abnormalities responsible for most cases.

Leukaemia, a group of blood diseases in which the white blood cells or their precursors become cancerous, and go into uncontrolled proliferation (the word 'leukaemia' comes from the Greek for 'white blood'). The white cells often show a chromosomal translocation, and it is likely that what happens is that a single white cell gets this chromosome damage which alters the genetic control of cell division (*see* p. 34); all the cancerous white cells are descendants of this one damaged cell (so they are clonal and have inherited the same damaged chromosome).

Not all the leukaemias have the same cause. Acute myeloid leukaemia seems to arise spontaneously and does not appear to run in families. Chronic lymphoid leukaemia, on the other hand, may have a genetic component, in that it has an association with >multiple sclerosis. Acute lymphoblastic leukaemia has a possible genetic link with the >HLA locus, but it has also been suggested that environmental factors are most important: this is the type of leukaemia that shows a high incidence near the Sellafield nuclear plant in the English county of Cumbria. *See also* Philadelphia chromosome.

Life cycle, in genetical terms, the cycle from fertilization through maturation to the fertilization of the next generation. It can also mean the stages through which any individual organism progresses, from fertilization to death.

Ligation, the joining together of separate sequences of DNA by DNA ligase enzymes. It occurs during replication (*see* >Okazaki fragments), DNA repair, and is also widely used in genetic engineering to stick DNA molecules together *in vitro*.

Line breeding, the technique of breeding from animals as closely related as possible, much used in improving agricultural stocks. It is not unusual for cows, for example, to be mated to their own sire or to a brother. *See* inbreeding.

Lineage, a sequence of species through evolutionary time, from the ancestral species to its present-day descendants.

L

Linkage, the tendency of genes that are on the same chromosome to be inherited together. Linkage of two loci provides an exception to Mendel's law of >independent assortment, which proposes that all genes are independent from one another. Clearly this cannot be so, given that genes are physically located in strings along chromosomes; those that are on the same chromosome are linked. To prove whether two given loci are linked, an F_1 cross is made between two strains, one breeding pure for the dominant alleles at each locus (*AABB*), the other for the recessives (*aabb*). The F_1 (*AaBb*) is then >backcrossed to the double-recessive strain, and if there is no linkage (i.e. the loci are on different chromosomes), there are four types of offspring in equal quantities: *AaBb*, *Aabb*, *aaBb*, *aabb*. If the loci are linked, however, the F_1 individuals will pass on only *AB* chromosomes or *ab* ones, so that in the offspring of the backcross, there are only *AaBb* and *aabb*. This is the result with very close linkage; but more usually there is a measurable proportion of cases in which linked alleles become separated from one another by the process of >crossing-over during meiosis.

Linkage group, a group of genes that can be observed to be linked together rather than to assort independently. In practice, a linkage group means a group of genes all on the same chromosome.

Linnaeus, Carl (also known as Carl von Linné) (1707–78), Swedish botanist. Linnaeus began a medical training but soon turned to botanical research. His first project was a system of >classification of flowering plants based on their sexual parts (*Systema naturae*, 1735). Economic necessity forced him to complete his medical training, and from 1741, he was professor of both medicine and botany at Uppsala university. He continued his work in classification, and perfected binomial nomenclature (*see* p. 13), the system under which all creatures have a genus name (noun) and a species name (adjective). His passion for classification blinded Linnaeus to the possibility that species could change, and he was sure that there had been no alteration in the living world since the Creation.

Locus (pl. 'loci'), the point on a >chromosome where a given gene is situated.

Lorenz, Konrad (1903–89), Austrian zoologist; pioneer in the study of animal behaviour. Lorenz trained as a doctor, but was prosperous enough to give up work and study bird behaviour. He believed that this could only validly be done in the wild (against the mainstream of psychological research at the time, which concentrated on reproducible experiments in the laboratory). Lorenz proposed that much behaviour is controlled by instinct, foreshadowing the line now taken by sociobiology. He discovered 'imprinting', the process whereby a bird learns what species it is from what creature it sees immediately after hatching (in many of Lorenz's experiments with geese, this creature proved to be Lorenz himself). Lorenz had >eugenicist ideas that fitted in well with those of the Nazis, whom he supported, although he came to regret his involvement with them. After the war, he continued his bird researches and also applied his ideas about instinct to human behaviour. His non-technical writings include *King Solomon's Ring* (1952) and the influential *On Aggression* (English edition 1966) (*see* aggression). Lorenz shared the 1973 Nobel prize for physiology or medicine with Nikolaas Tinbergen and Karl von Frisch, these three being the first people in the behavioural sciences to win the prize.

Lucy, the name given to the earliest known 'human' being, who lived in Ethiopia nearly 4 million years ago. *See* human evolution.

Luxury gene (a term rarely used nowadays), a gene that codes for a product specific to one particular type of cell, in contrast to >housekeeping genes. An example of a luxury gene would be the gene for insulin, which is active only in certain cells in the pancreas (though present in all cells). Luxury genes are switched off in all cells except the ones where their function is needed. While they are inactive, the DNA of luxury genes is highly condensed as >heterochromatin, and it shows up as a dark-staining band (*see* banding).

Lyell, Sir Charles (1797–1875), British geologist. Lyell took up the idea previously put forward by James Hutton (*see* p. 15) that the formation of the Earth had not been a single event but was a continuous process that was still in action and would continue for ever (*see* p. 12). He wrote the popular *Principles of Geology* (1830), which influenced Darwin's thought about evolution. Lyell was one of the friends who encouraged Darwin to publish *On the Origin of Species*, though he himself never accepted Darwin's theory.

Lyon hypothesis, the theory that, in female mammals (sex chromosomes XX), one X in each cell is inactivated so as to achieve >dosage compensation, the individual then having the same amount of X-chromosome activity as the male (XY). The X chromosome to be inactivated is chosen at random, and this happens in different tissues at different times during the development of the embryo (>embryogenesis). Once one X has been inactivated in one cell, all cells derived from that cell will have the same X inactivated. A good example of this is the >tortoiseshell cat (always female) which has patches of black and orange fur; the gene for fur colour is on the X chromosome, and the black patches derive from a cell in which the orange-gene X had been inactivated and vice versa. (In marsupials, X inactivation is not at random but always affects the X inherited from the father. There could never be a tortoiseshell wombat.) The female mammal is a naturally occurring >mosaic. The inactivated X chromosome can be seen microscopically as the >Barr body. The theory is called after the British geneticist Mary F. Lyon.

Lysenko, Trofim Denisovich (1898–1976), Soviet agricultural botanist. Lysenko was president of the Lenin Academy of the Agricultural Sciences from 1938, and also director of the Institute of Genetics at the Soviet Academy of Sciences from 1940 until 1945. He refused to accept the Mendelian genetics current in the rest of the scientific world, preferring the >Lamarckian system of the inheritance of acquired characteristics. His science was accepted as orthodoxy by Stalin (as it seemed to fit with a Marxist viewpoint), and Lysenko enjoyed great political patronage, though the great majority of Soviet scientists and notably >Vavilov knew he was wrong. After Stalin's death, criticism of Lysenkoism became possible, but he continued to occupy a highly influential position; the damage that he did to Soviet science and its international reputation has still not been repaired.

Lysis, the break-up of a cell after the rupture of its cell wall. This can be the result of attack by chemicals (e.g. detergents) or viral infection (as happens when bacteria are invaded by bacteriophages), or the cell may simply dissolve itself (autolysis). *See also* virus.

Lysogeny the colonization of a host cell by a >virus, which gets itself replicated by the host's genetic machinery.

M

MacArthur, Robert Helmer (1930–72), Canadian-born American ecological geneticist. Formerly a mathematician, MacArthur brought new theoretical insights to ecology. He proposed the theory of two different evolutionary strategies – opportunistic (*r* selection) and stable (*K* selection) – and gave the mathematical proof that both are compatible with Darwinian natural selection.

McClintock, Barbara (1902–90), American geneticist. McClintock's research into varieties of maize (sweetcorn) led, in the 1940s, to her discovery of jumping genes (now known as transposons), but the significance of these was not realized at the time. McClintock did further research on the role of jumping genes in the control of gene action, which was later developed by Monod. She belatedly won the Nobel prize for physiology or medicine in 1983.

Macroevolution, evolution viewed over a long time-span, especially concerning major trends. The development or disappearance of higher taxonomic groups is macroevolution. *See also* microevolution.

Major histocompatibilty complex, the group of genes that control the >HLA (human leucocyte antigen) system. *See also* immune system.

Malaria, an infectious disease in which the red blood cells are attacked by the protozoan parasite *Plasmodium*, which is transmitted from one infected person to another by *Anopheles* mosquitoes. Malaria is common in tropical areas, and used also to be common around the Mediterranean until mosquito numbers were reduced, partly through loss of their habitat as marshes were drained and partly by insecticidal campaigns. In those areas still affected by malaria, prevention is better than cure, and synthetic quinine-like drugs are effective in deterring the *Plasmodium* parasite.

Malaria, which causes high fever that often recurs for years after the original infection, is a dangerous disease, so it is not surprising that in areas where it is (or was) endemic genes that confer protection against it have been selected for and are common in the population. Unfortunately, in most cases these genes 'buy' protection against malaria at the cost of giving people who are >homozygous for the gene another disease: the two best examples are >sickle-cell anaemia and >thalassaemia, and another is >glucose-6-phosphate dehydrogenase deficiency (but one antigen in the >Duffy blood group series seems to give protection without deleterious effects).

Malthus, Thomas Robert (1766–1834), British clergyman and economist. Malthus graduated in mathematics from Cambridge University. His *Essay on the Principle of Population* (1798, 1803) stated that human populations increase in a geometric progression whereas the means for their subsistence (i.e. food) increase only in an arithmetic progression; and the factors that prevent an infinite increase of population are famine, pestilence and war. After the first edition brought a storm of controversy, Malthus produced a second five years later, containing a wealth of data from populations both historical and contemporary. In this edition Malthus suggested that 'moral restraint' (meaning late marriage) might also be a means of population control. His work had a seminal influence on >Darwin, and also on the economist David Ricardo. His work was also the first piece of serious research in demography, and is still a foundation of that science.

Manic-depression, a mental condition in which people are subject to extreme swings of mood not related to the events around them. In the depressive phase, they suffer wretched misery which may lead them to commit suicide. Manic-depression affects about 1 in 100 in European populations.

A breakthrough in the understanding of this disorder came through the study of the Amish, an American religious-fundamentalist community. In this small population there are quite frequent cases of manic-depression which run in families. An extraordinary finding was that all the 26 suicides among the Amish over the past 100 years had come from just four families. Analysis of the family pedigrees led to the conclusion that manic-depression is caused by a dominant gene, but that the gene has incomplete penetrance so that a person who inherits it has a chance (about a half to a third) of escaping the disease. By means of a genetic marker, the Amish manic-depression gene has been located on chromosome 11. Other researches into manic-depression have also found evidence of a dominant gene but without any connection with chromosome 11, so it is likely that there is more than one gene that can cause the same condition (as is suggested by its high frequency). There may also be an X-linked form. The mechanism of the gene's action is probably concerned with the neurotransmitters, molecules that are involved in communication between brain cells.

Map unit, a measure of the distance between two loci on the same chromosome. Loci are 1 map unit apart if cross-over occurs between them in 1 per cent of cases. A map unit has the formal name of >centimorgan.

Marfan's syndrome, an inherited condition with a number of effects, which are very variable: the person may be tall and thin, have extremely long limbs and digits, and have heart abnormalities. These various effects are the result of a single dominant gene, though the biochemical mode of action is not known (*see* pleiotropy). The symptoms do not cause any ill health until middle age, when the heart deformities often prove fatal; thus people with the syndrome have usually already had children by the time they are diagnosed, and the gene has been passed on to the next generation. About one person in 25,000 has Marfan's syndrome. It is said that the extraordinarily long fingers of the Russian composer and pianist Sergei Rachmaninov were caused by this syndrome.

Marriage. *See* monogamy.

Maternal effect, the situation in which the phenotype of the offspring is determined not by its own genotype but by its mother's. The classic example of this the coiling of the shell of the watersnail *Limnaea*. The alleles are *D* for right-handed coiling and *d* for left-handed, and are present in both sexes. To take a hypothetical pedigree, a *DD* snail (right-handed) has a *Dd* (right-handed) daughter who has a *dd* (right-handed) daughter who has a *Dd* (left-handed) daughter who has a *Dd* (right-handed) daughter. The system does obey Mendelian rules, but at a one-generation remove. There is some evidence to suggest that left-handedness in humans is inherited in this way. *See also* maternal inheritance; Sturtevant, Alfred Henry.

Maternal inheritance, any effect on the offspring that is not genetically transmitted but is passed on through the cytoplasm of the egg, i.e. >cytoplasmic inheritance. Maternal inheritance is the explanation of the situation in which a cross between a Strain A female and a Strain B male gives a different result from a cross between a Strain B female and a Strain A male (assuming that >X-linkage has been ruled out). *See also* maternal effect.

Mating system, the system within a population by which mates are chosen. At one extreme are parthenogenesis and self-fertilization; at the other is random mating. There may be systematic inbreeding or outbreeding, or assortative mating in respect of one or more traits. Some mating systems involve equal numbers of both sexes; in others, members of one sex have multiple mates. All mating systems have an effect on the genetic constitution of the succeeding generations: under random mating, it stays the same; inbreeding and positive assortative mating increase homozygosity; outbreeding and negative assortative mating increase heterozygosity.

Mean, the arithmetical average, i.e. the sum of all the values divided by the number of individuals. *See also* median; mode; normal distribution.

Median, the value in a distribution that comes half-way between the highest and the lowest. *See also* mean; mode; normal distribution.

Meiosis, the central event in heredity. It is the process of cell division that takes place when gametes (ovum or sperm) are made. The key feature is that the number of chromosomes is reduced from a set consisting of pairs to a set consisting of one copy of each. It works as follows:

First meiotic division

PROPHASE I After being elongated during >interphase (the non-division part of the cell cycle), the chromosomes begin to condense themselves (*leptotene stage*). The >homologous chromosomes come together and lie alongside each other (*zygotene stage*). The chromosomes double themselves lengthwise into two so that each consists of two >chromatids held together at the >centromere. There are now four chromatids lying alongside each other (*pachytene stage*), and the non-identical chromatids exchange sequences by crossing-over (*diplotene stage*). The chromatids continue to thicken and shorten (*diakinesis*).

METAPHASE I The membrane of the nucleus disappears and a fibrous structure known as the >spindle forms. The homologous pairs of chromosomes, now at their most condensed, orient themselves along the equator of the spindle.

ANAPHASE I The homologous chromosomes separate, the members of each pair moving towards opposite poles of the spindle (paternal and maternal chromosomes going either way at random).

TELOPHASE I The spindle disappears. Membranes for the two new nuclei may form.

Second meiotic division

PROPHASE II Each nucleus contains one member of each homologous pair, each consisting of two chromatids. The nuclear membranes disappear and spindles form.

METAPHASE II Chromosomes align themselves along the equators of the spindles.

ANAPHASE II The centromeres split, allowing the two chromatids of each chromosome to move to opposite poles of the spindle.

TELOPHASE II The spindles break down and membranes re-form round the four new nuclei; these each contain a set in which there is one representative of each chromosome (the erstwhile chromatids). If there has been any >crossing-over, which there usually has, each of the four versions of the chromosome is different.

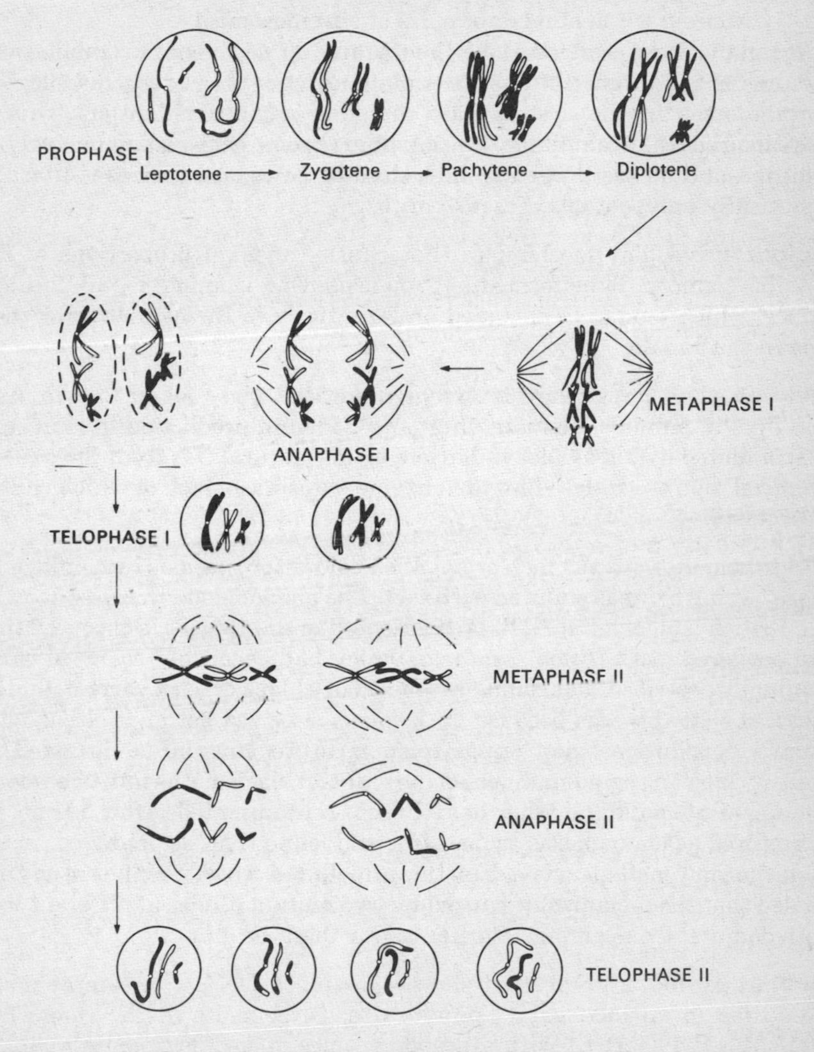

The process of meiosis is the same in all eukaryotes. It is also the same in males and females, but whereas the end result in males is four spermatozoa, the female only brings one egg to maturity, the other three products of meiosis being lost as >polar bodies. The whole process takes several days. The importance of meiosis in genetics is that it provides an endless source of variation over and above the variation contained in having a choice of alleles for most loci. Firstly, the maternal and paternal chromosomes are randomly divided at anaphase I. Even in a species with as few chromosomes as four pairs, this would mean only a 1 in 8 chance of an individual inheriting a complete set unchanged from those of a grandparent. However, for a human with 23 pairs, the chance is 1 in 4,194,304, and for a dog with 39 pairs, it is 1 in about 3 million (the formula is $1/2^{n-1}$ where n is the number of pairs of chromosomes).

Secondly, there is an immense opportunity for new genetic combinations to come about through the process of crossing-over between homologous chromosomes in the first meiotic division; *see* recombination. This is additional to the reshuffling due to whole chromosomes going one way or another, and means that each individual of a complex species is literally genetically unrepeatable. *See also* mitosis.

Meiotic drive, any mechanism that causes unequal proportions of the possible gametes to be formed at >meiosis. One example is >B chromosomes, which tend to be included preferentially in the ovum during meiosis in the female.

Melanin, the black or dark brown pigment that gives colour to skin, hair and eyes in animals, including humans. Melanin production in skin cells is stimulated by ultraviolet radiation, either natural (i.e. from the sun) or artificial, and is catalysed by the enzyme tyrosinase, lack of which causes >albinism.

Melanism, industrial, black or blackish coloration, used as camouflage by moths living in areas polluted with soot. The phenomenon was described by the British lepidopterist H. B. D. Kettlewell in the 1950s. He observed that the peppered moth *Biston betularia*, which had been light-coloured when originally described, was commonly found in a black or dark variant, and he deduced that this was because dark forms survived better in areas with industrial pollution, being less conspicuous to predatory birds. He tested his idea by releasing equal numbers of normal and black moths into two woods, one unpolluted and the other heavily soot-contaminated. After a time, he recaptured as many moths as possible, and found that, as predicted, many more normal moths survived in the unpolluted wood. Further study revealed that the melanism is caused by two mutant alleles at different loci, both dominant, one giving a darker colour than the other.

Melting profile, a graph that shows how much DNA in a sample melts over time in an increasing temperature. DNA melts in the range 60–80°C(140–175°F). DNA with a preponderance of A–T base-pairs melts at lower temperatures than DNA with more G–Cs, so the melting profile gives a crude indication of the composition of the sample.

Meme, a unit of cultural transmission (*see* exogenetic inheritance). The idea of a cultural unit of inheritance comparable to the physical unit (i.e. the gene) was put forward by the British zoologist Richard Dawkins in *The Selfish Gene* (1976). Its name is derived from the Greek *mimesis*, 'imitation', because the way in which a meme is transmitted from one person or group to another is by imitation. Dawkins' own examples of memes are 'tunes, ideas, catch-phrases, clothes, fashions, ways of making pots or of building arches'. He believes that memes are like genes in being able to replicate themselves, and in doing so for their own advantage rather than for the advantage of the individual carrying the meme. As an example of this, he cites the meme for belief in God, which propagates itself extremely readily, finding human culture a very comfortable environment. Humans are not the only species to have memes: two well-known examples from the animal kingdom are tits breaking milk-bottle tops and monkeys washing potatoes in the sea. The idea of memes is unacceptable to many geneticists and other scientists.

Mendel, Gregor (1822–84), Czechoslovakian (then Austrian) monk and botanist; discoverer of the laws of heredity. Mendel, a peasant's son, was interested in science from an early age and studied it at Olomouc before he joined the Augustinian monastery at Brünn (now Brno) when he was 21. His abbot sent him to the University of Vienna for two more years of scientific study (physics, chemistry, mathematics, zoology and botany), and on his return to the monastery in 1854, he taught mathematics and natural science at the local high school.

Mendel began his experiments in plant hybridization in 1856, and knew from the start that statistical results would be the key. He crossed over 20,000 plants of the edible pea and recorded the results. (Fisher noted that Mendel's data are just a little too good to be true; it seems possible that if in doubt he put a plant into the class that best fitted his theory.) Mendel revealed these results and propounded the simple two laws that bear his name – the law of >segregation (Mendel's First Law) and the law of >independent assortment (Mendel's Second Law) – and which are the foundation of all biology, in a paper which he read to the Natural History Society of Brno in 1865. It caused no stir either at the meeting or when it was published in the Society's journal the following year. Mendel was disappointed, but hoped that further research would bring more evidence so that his theory would be accepted. He experimented on other plants for the next two years until he was elected abbot of his monastery in 1868. After that his administrative duties took up all his time, and he died a respected abbot but an unrecognized scientist.

Sixteen years later, his name became known to the world when his theory was independently arrived at by three scientists – de >Vries, >Correns and >Tschermak – all of whom found Mendel's paper of 1866 and publicized his findings.

Mendelize, a useful verb which according to the *Oxford English Dictionary* means 'to exhibit or transmit parental characters in conformity with Mendel's law of inheritance; to behave as Mendel's law predicts'. Apparently first used in 1906, it has now totally disappeared, which is a pity.

Menstruation, a physiological arrangement unique to human beings and the great apes. In all other animals there is a periodic breeding cycle controlled by hormones, so that there is a regulated time during which sexual activity occurs, the female body only being fertile and in the right physiological state to begin and maintain pregnancy at that time (>oestrus in mammals is an example). The system of menstruation is quite different, in that it enables females to be sexually receptive all the time.

The change from oestrus to menstruation is probably a very significant evolutionary change, enabling different social structures to emerge in the higher primates. The British zoologist Sir Gavin de Beer considered that it was a necessary precondition for the evolution of >monogamy. (De Beer is almost the only biologist to have paid any attention to menstruation. It appears to have been entirely ignored by sociobiologists, even though it is a clear case of a physiological system having behavioural consequences, just the kind of topic that sociobiology deals with.)

Mental illness is distinguished from mental handicap in that the latter involves retarded development or brain damage whereas the former involves inappropriate reactions on the part of a person with a normal mental development. Degenerative diseases, e.g. >Alzheimer's, are usually included under mental illness.

Certain forms of mental illness have a genetic component in their causation, though it is never as straightforward as a gene 'for' disease X – there are always environmental factors too. *See* manic-depression; schizophrenia.

Meristic variation, variation in the phenotype that goes by step, e.g. the number of fingers. Such variation may have the same sort of underlying genetic cause as >continuous variation, or there may be a single gene responsible (e.g. a gene in mice that reduces the number of toes on the front feet from four to three). Meristic variation includes such traits as the number of puppies in a litter.

Meselson, Matthew Stanley (1930–), American molecular biologist; professor of natural sciences at Harvard University since 1976. In 1957 Meselson and his colleague F. W. Stahl were the first to confirm Crick and Watson's suggestion that when DNA replicates it does so by unwinding down the middle so that each of the two new molecules contains one strand of the original and one new one (i.e. semi-conservative replication). Their technique involved the use of radioactively labelled samples of DNA in >*Escherichia coli*.

Messenger RNA (mRNA), the molecule that transcribes the genetic message from DNA at the start of >protein synthesis. In prokaryotes, mRNA reads directly off the DNA and has its front end already bound to a ribosome making the protein while, further down, the molecule transcription is still taking place. In eukaryotes, there is a precursor of mRNA which is called heterogeneous nuclear RNA, and it is this that actually reads the DNA. Then all the sequences on it that represent introns (non-informational DNA) are cut out, the result being an mRNA molecule containing only sequences that represent exons (coding sequences). Eukaryotic mRNA is synthesized in the cell nucleus, where it reads the DNA, but it must travel out of the nucleus for the protein to be assembled on the ribosomes. After protein synthesis, the mRNA molecule may be reused for another translation or its constituent molecules may migrate back into the nucleus to be incorporated into another molecule of mRNA. The front (5′) end of an mRNA molecule has a >cap, and the 3′ end has a poly-A tail.

Metacentric, a chromosome having its >centromere at or near the middle.

Metafemale, a sterile female *Drosophila*, which has two sets of autosomes and three X chromosomes. It provides evidence for the role of the balance between autosomes and sex chromosomes in >sex determination.

Metamale, a sterile male *Drosophila*, which has three sets of autosomes and one X chromosome. It provides evidence for the role of the Y chromosome in >sex determination.

Metaphase, one of the phases of the two types of cell division, >meiosis and >mitosis. *See* diagram at meiosis.

Microevolution, evolution viewed over a relatively short time-span, and involving changes of gene frequency within populations. It can also concern the process of speciation, but not the development of higher taxonomic groups (*see* macroevolution). For an example of a microevolutionary study, *see* melanism, industrial.

Migration, movement of populations, groups or individuals. In genetic terms, migration inwards (immigration) is an important factor in maintaining genetic variability within a population's gene pool. Outwards migration (emigration) is a mechanism of population control, and can lead to the foundation of new populations.

Milstein, Cesar, (1927–), Argentinian-born British molecular biologist. Milstein taught chemistry in Buenos Aires before joining the Medical Research Council's Molecular Biology Unit at Cambridge, England, in 1963. His work on immunoglobulins (with Fred >Sanger) led to the discovery in 1975 of monoclonal antibodies, the single pure antibodies produced by a clone of >B lymphocytes (*see* immune system). Milstein shared the 1984 Nobel prize for physiology or medicine with his colleague George Koehler and the theoretician Niels Jerne who, following Milstein's results, developed the clonal selection theory of immunity.

Mimicry, the process which one species evolves so as to resemble another. It is found especially in butterflies that are preyed upon by birds.

Many such species have evolved unpalatability to deter predators, but

this system costs a certain loss of individuals as each new generation of predators has to learn the hard way (i.e. by attempting to eat one) that the members of a particular species are inedible. Mimicry can make unpalatability more efficient, and this can be done in one of two ways. In *Mullerian mimicry*, two or more unpalatable species come to look like each other, the advantage to each species being that a proportion of the sacrifice involved in educating, say, young birds is shared with the other species. The more alike the species look, and the more uniform the individuals within each species, the better it works. Conversely, *Batesian mimicry* is the system in which a species that is edible comes to resemble one that is unpalatable. This only works if the mimetic (mimicking) species is substantially less common than the unpalatable one; otherwise the predators would not learn that butterflies with that pattern are inedible. Some species that practise Batesian mimicry have several different mimetic forms, each an imitation of a different unpalatable species, so as to keep the numbers of mimics suitably low. The genetic control of mimetic forms is often by means of a >supergene at a single locus, sometimes with modifier alleles at other loci.

Mismatch repair, enzymatic removal from a DNA molecule of a base that is an incorrect match with the one opposite, and therefore is not properly bonded. It is replaced with the right one. *See also* >DNA repair.

Missense mutation, a mutation in which a codon that codes for one particular amino acid is changed to a codon that codes for another, with a consequent change in the chemistry of the gene product. *See* genetic code.

Mitochondria (sing. mitochondrion), >organelles in the cytoplasm of >eukaryote cells that are concerned with respiration (generation of chemical energy by oxidation). There are on average a few hundred mitochondria in each cell. They carry their own DNA, which can become the vehicle of >cytoplasmic inheritance. The base sequence of human mitochondrial DNA has been deciphered in its entirety.

Mitosis, the process of ordinary cell division. The object is to produce two daughter cells each with exactly the same chromosome complement as the original cell (in contrast to >meiosis, the other type of cell division, which produces >gametes). The chromosomes behave in the following way:
INTERPHASE This is the non-division part of the cell cycle, but towards the end of it the chromosomes, still in elongated shape, duplicate themselves lengthwise into two >chromatids (*S phase*; *see* interphase).
PROPHASE The chromosomes begin to condense themselves by coiling up tightly. The nuclear membrane disappears. The fibrous structure known as the >spindle is formed, and each chromosome attaches itself to a spindle fibre by its >centromere.
METAPHASE All the chromosomes move towards the equator of the spindle. This is the time at which the chromosomes are at their most condensed and easiest to observe.

ANAPHASE The centromeres split, allowing the two sister chromatids to move to opposite poles of the spindle. Each chromatid will consitute a chromosome in one of the new cells.

TELOPHASE Nuclear membranes form round each of the two groups of chromosomes. The spindle disappears. The chromosomes elongate, and get on with whatever task of protein synthesis is required in the new cells.

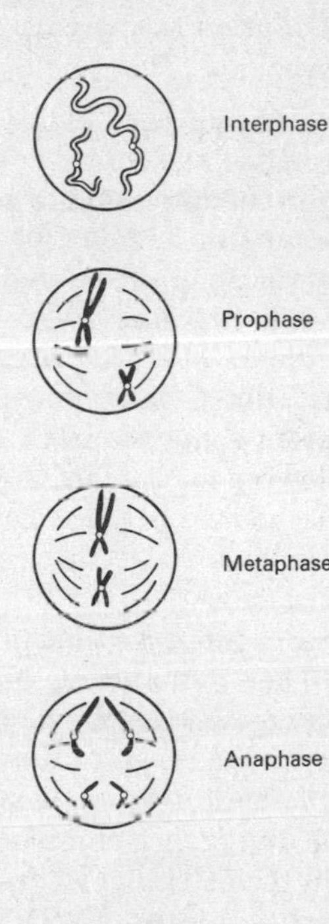

MLD, 'mean lethal dose', the amount of whole-body radiation that, given in a single dose, will kill 50 per cent of the individuals receiving it. For humans, the MLD is believed to be about 450 rem.

MN blood group system, a series of genetically determined antigens found on human red blood cells. The second blood group system to be discovered, it was found in 1927 by the blood-group pioneer Landsteiner and his colleague Levine, who had discovered the ABO system in 1900. The M and N antigens are so alike that each is recognized as 'self' by the body's immune system whether it is the individual's own type or not. It therefore is not necessary for blood transfusions to match the MN type of donor and recipient. The genes for M and N are about equally common in Europe, so that the population consists of approximately 25 per cent group M (homozygotes), 50 per cent group MN (heterozygotes) and 25 per cent group N (homozygotes). Australian Aborigines have frequencies of M of 75–95 per cent.

Mode, the value in a distribution that occurs most often. *See also* mean; median; normal distribution.

Molecular clock. *See* molecular evolution.

Molecular evolution, evolutionary change of particular molecules. There are many proteins that occur in a similar form in many different living things, and the evolutionary path of such a molecule from its early form to its later ones can be traced. In cases where the DNA itself has been sequenced for a gene that occurs in a wide range of creatures, the number of mutations (base substitutions, deletions, insertions) can be directly compared. If only the protein sequence is available, the number of mutations can be deduced, allowing for the fact that a proportion of mutations are silent (i.e. the same amino acid is coded for even after a base change). One must also allow for the fact that sometimes a second mutation happens on the site of an earlier one.

The 'molecular clock' model proposes that mutations happen to a molecule at a particular rate per generation, and that by comparing the number of mutational changes between two species, one can deduce the number of generations (hence years) since they were the same. A study of the alpha and beta globins found that molecular clock time agreed well with geological time in the divergence of major evolutionary groups (four-legged creatures from fish, terrestrial creatures from amphibians, warm-blooded animals from reptiles and so on). Different proteins have molecular clocks running at different rates, and some parts of the molecule will be more liable to change than others (with some sequences completely invariant). For example, cytochrome *c* and histone H4 are extremely conservative, there having been only four mutational changes in the latter between flowering plants and mammals. On the other hand, very rapid change occurs in the satellite DNA sequences in *Drosophila*, enabling the evolutionary divergence of the hundreds of species of this fruit fly to be charted over a time-span as short as one million years.

Monera, one of the five kingdoms of living things (*see* classification). It includes all the >prokaryotes.

Monoclonal antibodies, a synthetic preparation of pure antibodies. The technique of producing monoclonal antibodies was devised by the British molecular biologist Cesar >Milstein in 1975. Given that each line of >B lymphocytes only produces one type of antibody, if a particular antibody is wanted the procedure is to inject a mouse with the appropriate antigen, and wait until the clone of lymph cells in its spleen begins to produce the antibody. Some of these cells are extracted, and instead of simply being cultured (lymphocytes will not grow indefinitely *in vitro*), they are fused with myeloma (cancerous) lymphocyte cells that proliferate speedily. These hybrid cells (known as hybridomas) combine the useful properties of both parental cells – an ability to produce the required antibody, immortality and rapid proliferation. Antibodies can be produced in quantity by this technique, for use in tissue typing and blood typing before transplants and transfusions. Monoclonal antibody is also used in the treatment of certain cancers, and to prevent haemolytic disease of the newborn of Rhesus-negative mothers.

Monod, Jacques Lucien (1910–76), French molecular biologist; co-proposer of the theory of the operon. Monod was trained as a zoologist, and worked at the Pasteur Institute in Paris. Immediately after the discovery of >DNA's structure, he and his colleague François >Jacob were stimulated to look for the mechanism that controls whether a length of DNA – a gene – is active or not. In 1961 they proposed the mechanism that they called the operon, and also predicted the existence of the RNA molecule now known as messenger RNA. For this work, Monod and Jacob shared the 1965 Nobel prize for physiology or medicine with André Lwoff (for his work on >bacteriophages). Monod also wrote on the philosophy of evolution, proposing in his *Chance and Necessity* (1970) that the origin of life was a matter of chance but that its subsequent evolutionary path has been a matter of necessity.

Monoecious, adjective used to describe a plant species which has separate male and female flowers borne on the same plant. *See also* dioecious; flowering plants.

Monogamy, the mating system in which one male and one female pair exclusively, either for a season or for life. It is rare, being found in only 3–5 per cent of mammal species and a few fish, but it is the usual state for birds, with about 90 per cent of bird species being monogamous; it is hardly ever found in other vertebrates or in invertebrates.

In most species, the male's best chance of increasing the number of his genes that are passed on to the next generation is by mating with as many females as possible. But in species in which the young need more care than can be provided by one parent, it pays the male to stay with one female and raise a brood of young successfully rather than to sire a number of broods that will die. Typically, fledgling birds need one parent to keep them warm while the other forages for food, and the male and female have a similar anatomy that makes them able to share these tasks. Another cause of monogamy among bird and other species is a habitat in which only a low population density is possible, in which case it is not feasible for a male to control more than one female as they would be too widely scattered.

In most mammalian species, care of the young can be done by the mother alone, as she feeds them with her milk. The monogamous mammals are mostly canids (members of the dog family) having large litters which require the mother to be dependent for a long time on food brought to her by a mate. The young still have to have food brought to them by their parents even after they are weaned; and canids have evolved the ability to swallow food to carry it back to the lair and regurgitate it for the young. Among the higher primates, the gorilla and the chimpanzee are not monogamous; the orang-utan and the gibbon are, probably by reason of population density. As for humans, some societies are monogamous but more are not.

Monogenic, adjective describing a trait that is controlled by a single gene. This would include most of the human clinical syndromes.

Monohybrid, an individual that is heterozygous at a single locus. The term is also used for a cross involving two different alleles at a single locus, equivalent to an >F_1.

Monomorphic locus, a locus at which the most common allele has a frequency of 95 per cent or greater. (Strictly it should mean a locus at which only one allele is present in the population, but mutation will always occur even at a low rate, thus providing alternative alleles.) Studies of natural populations of *Drosophila* indicate that as many as 90 per cent of loci are monomorphic, and this is probably typical. *See also* polymorphism.

Monophyletic group, a group of species (or higher taxa) descended from a common ancestor. Humans and the anthropoid apes are a monophyletic group, all being descended from a common ape-like stock. *See also* polyphyletic group.

Monosomy, the condition of having only one representative of a pair of chromosomes. Monosomy can cause abnormalities, of which the best known is Turner's syndrome, monosomy for chromosome X in humans.

Monozygotic twins, identical twins, those who are genetically identical because they come from a single fertilized egg which has divided in two. It is interesting that monozygotic twins are happier about being twins than dizygotic (fraternal) twins are, with one poll revealing that 78 per cent of the former are pleased to be twins as against 42 per cent of the latter. However, the idea of an unfailing mystical bond between identical twins is not borne out by the fact that 10 per cent of them dislike being twins (the data from which these figures come do not say whether these 10 per cent of twins are from the same pairs.) Monozygotic twins tend to 'mirror' each other; for example if one twin is right-handed the other is left-handed. Monozygotic twins are much studied by geneticists, as they offer the most convenient way of estimating the heritability of a trait (*see* twin studies). *See also* polyembryony, twinning rate.

morgan. *See* centimorgan.

Morgan, Thomas Hunt (1866–1945), American geneticist. Following William Sutton's suggestion, in 1902, that chromosomes might be something to do with Mendel's 'factors', Morgan experimented ón *Drosophila melanogaster* and found that the 'factors' are indeed located on the chromosomes and that they are arranged linearly. Morgan's 'fly room' at the California Institute of Genetics became a genetical kitchen unequalled until the great days of the Medical Research Council's Molecular Biology Unit at Cambridge. In 1911 Morgan and his colleagues (including A. H. >Sturtevant) produced the first-ever genetic map, showing the location of five X-linked genes in *Drosophila*. Morgan was a skilled writer about genetics for the lay reader, and in his book *The Scientific Basis of Evolution* (1932) he wrote about the unique capacity of humans to evolve by non-biological means; his ideas are very close to those now proposed by Richard Dawkins (*see* meme). Morgan was the first geneticist to win the Nobel prize for physiology or medicine (1933).

Morula (from the Latin *morus*, 'mulberry'), the mammalian embryo when it consists of a ball of undifferentiated cells. *See* embryogenesis.

Mosaic, an organism whose body contains cells of two (or more) different genetic constitutions, though all of them derive from the same gamete. This could happen as the result of a mutation in a somatic cell, so that some cells have the normal genotype but the cells in the line derived from the one with the mutation are different. A particular sort of mosaic occurs in female mammals, as the result of random inactivation of one of the two X chromosomes (*see* Lyon hypothesis). *See also* chimaera.

Mouse. *See Mus musculus*.

mRNA. *See* messenger RNA.

Mule, the offspring of a male ass or donkey and a female horse or pony. The reverse is a >hinny, though it is often referred to as a mule, there being no significant physical difference between them. The diploid chromosome number in the ass or donkey (*Equus asinus*) is 62, and in the horse or pony (*Equus caballus*), it is 64, so the mule has 63 chromosomes, and is infertile.

Muller, Hermann Joseph (1890–1967), American geneticist. Muller worked in T. H. >Morgan's 'fly room', concentrating from the start on mutation. In 1920 he expounded the theory that, as only genes can reproduce themselves, all other components, even down to the fine structure of cells, must ultimately be produced by the genes. From this he concluded that life itself must have started with simple self-replicating molecules, which he termed 'naked genes'. As good an experimentalist as he was a theorist, Muller discovered in 1926 that X-rays induced a high rate of mutation, and he suggested (correctly) that a mutation is a chemical event within a gene.

Muller was concerned with eugenics, and was an advocate of sperm banks stocked with contributions from distinguished men (see p. 42). Muller won the 1946 Nobel prize for physiology or medicine.

M

Multigene family, a set of similar genes that code for different parts of a particular function. They presumably arise by duplication and then diverge slightly so that they each contribute their different products. An example in humans is the group of genes coding for the different chains of the haemoglobin protein.

Multiple alleles, the situation in which a population has more than two alleles regularly occurring at a given locus. *See* polymorphism.

Multiple births, in humans, twins (*see* dizygotic twins; monozygotic twins; twinning rates), >triplets, >quadruplets, quintuplets, and so on. These occur naturally (there being some genetic element in the tendency to have multiple births), but have increased ten-fold in Western countries as a result of treatment with a fertility drug which causes more than one egg to ripen in each monthly cycle. It is given to women whose ovulation is poor, and often results in several eggs being fertilized at once.

More importantly, it is given to women as part of *in vitro* fertilization treatment, so that several embryos at once can be implanted to increase the chance of pregnancy. The drawback is that multiple babies are bound to be born premature and to have a poor chance of survival. One way round this is to selectively abort surplus foetuses once the pregnancy is established, but this is very risky and can result in the loss of all the foetuses; it is in any case stressful for the mother. A better solution is never to implant more than three or four embryos, a number that can have some hope of survival. Even so, there has been criticism that the multiple babies produced artificially by *in vitro* fertilization treatment are taking up scarce resources in neonatal intensive-care units that are needed for naturally occurring premature babies.

Multiple sclerosis, a disease in which the nervous system degenerates, causing loss of co-ordination, paralysis and early death. Its causes are not known, but there is some genetic predisposition in that the majority of sufferers are of HLA type DR2. Environmental factors are more important, however, as shown by the very different frequency in white populations (of similar European ancestry) in different parts of the world: the disease has its highest incidence in northern Europe, especially in Britain and Scandinavia, but is uncommon in North America and rare in Australia and New Zealand.

Multistate character, a genetically determined trait that can exist in three or more discrete states. An example is the >ABO blood group system. *See also* binary character.

Mus musculus, the house mouse; the mammal most often used in genetical research in the laboratory. Mice are found all over the world, both in the wild and in populations supported by humans, in which case they are a serious economic pest. A mouse is about 15 cm long (including the tail), weighs 25–30 g, and is omnivorous. Gestation time is 19–20 days, and the average litter is 4–6 pups (range 1–14), which mature within about two months; the lifespan is three years. Wild mice are almost always >agouti coloured, but different colours have been bred since ancient times in China and Japan and are found in some laboratory strains (with still more varieties among the mice bred as a hobby). There are over 50 main strains of mouse whose records go back to the 1920s (a proof that inbreeding is not necessarily bad). The mouse's diploid chromosome number is 40. Many mutant genes have been identified in mice, including a number that cause the same diseases as in humans, e.g. leukaemia and diabetes. *See also* nude mice; tail-less; waltzing mice.

Muscular dystrophy. *See* Duchenne muscular dystrophy.

Mutagen, any substance that raises the frequency with which >mutations occur. It may be a form of radiation, e.g. ultraviolet light or >X-rays (whose mutagenic effect was discovered in 1926 by H. J. >Muller), or a chemical, e.g. mustard gas.

Mutation, any change occurring in the genetic message that an >allele carries. The first theoretician of mutation was H. J. >Muller, who saw that the three important factors are: *(1)* that Darwinian evolution requires a constant supply of variation, much of it supplied by mutation; *(2)* that mutations are extremely rare; and *(3)* that mutations are almost always deleterious. It was later proved, as evolutionists had supposed, that 'mutations do not arise in response to an organism's needs nor do they, except by accident, gratify them' (the discovery was Luria and Delbruck's, the words are Sir Peter Medawar's). A mutation is likely to be deleterious because it is a random change to a message – and therefore a molecule – that already works well. There are three general types of mutation: those involving a substitution at a single base-pair (*see* point mutation; missense mutation; nonsense mutation); those involving the deletion or insertion of one or a few base-pairs (usually resulting in a >frameshift); and large-scale rearrangments of chromosomal material. A mutation that is a change from the normal allele to a new form (which is the case in the vast majority of mutations) is called a >forward mutation; these are occasionally reversed by >back mutations. *See also* mutation rate; neutral allele.

Mutation load, the total genetic burden carried by a population that results from accumulated deleterious mutations.

Mutation pressure, the effect of recurrent mutation in changing the gene frequency at a given locus. This pressure is usually very weak compared to selection pressure or the effect of genetic drift.

Mutation rate, the rate at which mutational changes in DNA occur. It is expressed as the probability that an allele will undergo mutation in a single generation, or in each gamete formed (both measures are used). It was first studied by the American geneticist H. J. >Muller, who realized that the rate of mutation in nature is very low, and discovered that it could be enormously speeded up by doses of X-rays and other mutagens. Different genes have different mutation rates: the range in *Drosophila* is from 10^{-5} to 10^{-6} per gene per generation, or 10^{-8} to 10^{-9} per base-pair per generation. Similar rates are found in humans, whose mutation rate in respect of single-gene medical disorders can be estimated (for example, >achondroplasia arises at a rate of 1.3×10^{-5} per generation). There are some mutations that themselves influence the rate of mutation of other genes around them, and it is likely that mutation rates are themselves the result of genetic control and have evolved adaptively.

Myxoviruses, a group of viruses that infect vertebrates. The single-stranded DNA genome is encased in a spiral protein coat. Among the diseases caused by myxoviruses are influenza, mumps, measles, rubella (German measles) and rabies. (Myxomatosis, the disease that is lethal to rabbits, is caused by one of the >poxviruses.) These viruses tend to undergo frequent genetic change, so that vaccines that worked against previous versions of the virus are ineffective against a new strain.

N

Nail–patella syndrome, a condition in which the nails of fingers and toes are deformed and the patellae (knee-caps) are very small; there is no serious effect on health. The syndrome is caused by a single dominant gene, but the biochemical mechanism is unknown. The interest of the condition for geneticists is that the nail–patella locus is close to that of the ABO blood group system, and some of the earliest studies on human autosomal linkage were carried out using these two sets of genes.

Natural immunity, the system of immunity that is non-specific, i.e. does not rely upon recognition of antigens. This system works by *(a)* production of the anti-viral interferon proteins, *(b)* lysozyme activity to attack bacteria, *(c)* ingestion of particles, and *(d)* the action of various proteins to agglutinate foreign matter (i.e. to make it stick together in clumps). This system is the only one used by invertebrate animals; vertebrates have evolved a much more elaborate >immune system but still use natural immunity alongside.

Natural selection, the mechanism by which evolution happens. Natural selection is simply the process which ensures that some genes are relatively more numerous in the next generation than in the present one, because of the reproductive success of the individuals carrying these genes. Spencer's catch-phrase 'survival of the fittest', used to describe the action of natural selection, is misleading in two ways: firstly, 'fittest' means 'best adapted' not 'in the best condition'; and secondly, survival in itself is not necessarily important. The crucial factor is differential fertility: two organisms might live the same length of time (i.e. both are survivors), but if one raises a hundred offspring and the other only eighty the former has been favoured by natural selection. It is also important only to consider *successful* reproduction, i.e. the production of offspring who themselves live to reproduce. There is no profit, evolutionarily speaking, in putting one's genes into large numbers of offspring who do not pass them on.

The term 'natural selection' has, as Darwin was aware, an unfortunate implication that nature makes choices; this is clearly not so, 'natural selection' being simply a description of what happens to gene frequencies. Natural selection is also not purposive: the outcome may be either 'good' or 'bad' for the species involved in the long run; all that is being selected for is adaptedness (as measured by reproductive success) in the present environment. (*See* albinism for an interesting example.)

Different types of natural selection have been identified: *(a)* acting in different directions (*see* directional selection; disruptive selection; stabilizing selection); *(b)* favouring different reproductive strategies (*see K* selection; *r* selection); *(c)* acting other than upon individuals (*see* group selection; kin selection). *See also* sexual selection.

Nature versus nurture, the never-ending debate on whether >genotype is more important than environment in human development and, in particular, in determining intelligence. The term was coined by Galton, who took a great interest in the matter and wrote several books on it, one entitled *English Men of Science: Their Nature and Nurture*. It is naïve to think of any trait as having a fixed proportion contributed by nature, the rest being contributed by nurture; see the examples under >heritability. Rather, it is an axiom of genetics that all variation is the result of interaction between genotype and environment (this is not to say that one or the other may not in fact contribute 100 per cent in some circumstances). None the less, the political implications of the nature-versus-nature debate in terms of the development of intelligence are so important that there have been repeated attempts to 'prove' that differences in intelligence are either due to the environment (and therefore removable) or to genotype (and therefore inevitable). Evidence such as Galton first gathered, that intellectual ability runs in families, is equally supportive of the environmentalist or the hereditarian position; but Galton was also the first to realize that the study of monozygotic (identical) twins could be informative *(see* twin studies). *See* IQ for more about that issue, and *see* >adoption for the evidence from adopted children. Nature versus nurture has also been debated in connection with other continuously varying human traits: temperament, behaviour (notoriously, criminality) and susceptibility to certain diseases (*see* obesity).

- Lancelot Hogben, *Science for the Citizen* (1938; unfortunately out of print). Contains an excellent discussion of nature versus nurture, which although written more than 50 years ago says all that needs to be said.

Neanderthal Man. *See* human evolution.

Negative control, the situation in which a gene is not transcribed until the repressor molecule is dissociated from the operator region of the

>operon (for inducible enzymes). In the case of negative control of repressible enzymes, the gene is normally transcribed until repressed. The opposite is >positive control.

Neo-Darwinism, the synthesis of Darwinian evolution and Mendelian inheritance, as put together in the 1920s and 1930s by the great geneticists >Fisher, Sewall >Wright, and others (*see* pp. 26–7). Darwin had not understood the mechanism of heredity, and his theory was weakened by lack of it; Mendel had not known about evolution by natural selection. Neo-Darwinism unites the two, and is the basis of all subsequent thinking in biology. It is not to be confused with >social Darwinism.

Neoteny, the retention in adult life of structures or stages of development that were appropriate for an immature stage in the ancestral form. A striking example is the axolotl, a Mexican salamander, which reaches sexual maturity while still in the larval state. *See also* paedomorphosis.

Neurofibromatosis, an inherited disease in which there are tumours in the nerve cells, first in the peripheral nerves and later in the spine and brain. Onset is at about the age of 20, and death at about 50. The disease is caused by a dominant gene, and occurs in about 1 per 2500 babies in European populations.

Neurospora crassa, a fungus, known as the pink bread mould, and often used in genetic experiments. For much of its life cycle, *Neurospora* grows as filaments called >hyphae, which are of two 'mating types'. To reproduce, the hyphae (which are haploid) produce >conidia (also haploid), which are spores that fertilize the opposite mating type. The zygote so produced is diploid and undergoes meiosis to produce four new haploid cells; each of these then undergoes mitosis, and the resulting eight haploid cells (called ascospores) are arranged in a row in the ascus, like peas in a pod. They can be removed and each cultured separately to be analysed genetically. This gives a very convenient way of studying cross-over frequency. As *Neurospora* is haploid in its hyphae state, all recessive alleles are expressed; this fact helped George >Beadle and E. L. >Tatum to establish that a single mutant causes the loss of a single enzyme.

Neutral allele (or neutral mutation), an allele that is a variant from the normal but does not give any selective advantage or disadvantage to the organism that carries it. According to some theoreticians, such alleles are numerous and can spread widely within a population, as there is no selection against them; others believe that it is hardly possible for a mutation to be truly neutral in selective terms. *See* polymorphism.

Neutralists. *See* genetic drift.

Nexus hypothesis, the idea that every phenotypic trait is probably influenced by more than one gene, and that each gene has an influence on more than one phenotypic trait. It is not, as appears at first sight, a contradiction of the axiom of 'one gene, one enzyme' (*see* pp. 27 and 29).

Nick, a cut in a single strand of a double-stranded DNA molecule. *See also* restriction endonuclease.

Nilsson-Ehle, Herman (1873–1949), Swedish geneticist. A farmer's son, Nilsson-Ehle was (as Mendel had been) interested in applying science to improving crops. Realizing the significance of Mendel's work, he proved that the economically important traits in plants – size, earliness, resistance to disease – were Mendelian, even though they were continuously varying where Mendel's traits had been discrete. His papers on this (1908–11) are classics of genetics.

Nirenberg, Marshall Warren (1927–), American molecular biologist; has worked for the US National Institutes of Health since 1957. Nirenberg was the first person to decipher a 'word' of the >genetic code. In 1961, he synthesized a molecule of messenger RNA consisting of the base U (uracil) repeated indefinitely. The resulting protein consisted of nothing but the amino acid phenylalanine over and over again, so he could deduce that the codon UUU codes for phenylalanine. For this, Nirenberg shared the 1968 Nobel prize for physiology or medicine with Har Gobind >Khorana and Robert W. Holley.

Non-disjunction, the failure of homologous chromosomes to separate during >meiosis, or of sister chromatids to separate during >mitosis. The result is that the descendant cells after this event have an uneven amount of genetic material in them: too much in one line, too little in the other. Non-disjunction can give rise to >trisomy, the condition in which there are three chromosomes in a 'pair' instead of two, as in Down's syndrome.

Non-homologous chromosomes, any chromosomes that do not have an exactly matching pair. The main example is the sex chromosomes, in which the X and Y are non-homologous.

Non-identical twins. *See* dizygotic twins.

Nonsense codon (also known as stop codon), one of the three codons that specify the end of a genetic message. These codons do not code for any amino acid but signal the termination of the protein being synthesized. They are UAG ('amber'), UAA ('ochre') and UGA ('opal'). *See* genetic code; protein synthesis.

Nonsense mutation, a mutation in which a codon that codes for an amino acid is changed to a codon that codes for 'stop' (*see* nonsense codon). The result is a shortened protein, how much shorter depending on how far along the gene the nonsense mutation occurs. *See* genetic code; protein synthesis.

Normal distribution, or Gaussian distribution, the statistical distribution in which the frequency of occurrences of values decreases with their distance from the mean. When plotted as a graph, a normal distribution looks symmetrically bell-shaped. The mean (the average of all the individual values), the mode (the value that occurs most often) and the median (the half-way point between the highest and the lowest values) all coincide. The >standard deviation is a measure of how steep-sided the 'bell' is. In biological data, many traits that show continuous variation conform to a normal distribution. *See also* bimodal distribution.

N

Northern blotting, a method of identifying sequences of RNA, by a technique similar to >Southern blotting (which is used for DNA). Whereas Southern blotting was called after its inventor, Northern blotting got its name by analogy.

Nucleic acid, one of the polymeric molecules consisting of nitrogenous bases connected to a sugar–phosphate structure. They can be either double- or single-stranded. The ones that occur in nature are >DNA and >RNA, and these can be artificially joined by hybridization (*see* Hybrid DNA).

Nucleolus, a spherical organelle within the cell nucleus where >ribosomes are made. There are about five or six in each somatic cell nucleus.

Nucleoside, a compound consisting of a nitrogenous base attached to a sugar. *See also* nucleotide.

Nucleosome, the structural unit from which chromosomes are made up. Each nucleosome consists of a block of eight molecules of the protein >histone with a length of about 145 base-pairs of DNA wound round it, plus a linker unit of DNA of a length of about 55 base-pairs, and a single histone molecule to join it to the next nucleosome.

Nucleotide, a >nucleoside (a compound consisting of a base joined to a sugar) with a phosphate group attached. Nucleotides are the units of which the nucleic acids DNA and RNA are composed.

Nucleus, the roughly spherical structure in a eukaryote cell that contains the chromosomes. It is also the place where messenger RNA is made (by transcription of the DNA in the chromosomes); these RNA molecules must pass out through the nuclear membrane to the cell's cytoplasm where the ribosomes carry out protein synthesis.

Nude mice, mice that are homozygous for the allele *nu*. These mice have no thymus gland, a crucial organ in the immune system as it produces T lymphocytes, the cells that recognize foreign particles or tissue. These mice are therefore very susceptible to any kind of infection, but they will accept grafts or transplants from any donor even if genetically completely unrelated. Thus they are often used in experiments concerned with transplant rejection. The fact that they are hairless is a side-effect of the same gene (*see* pleiotropy).

Null hypothesis, the assumption that the effect being statistically investigated does not exist. For example, if one has two samples with different means for a particular trait, the null hypothesis is that they actually come from the same population. Only if the means can be shown to be different at a level that is statistically significant can the null hypothesis be rejected and the samples accepted as (probably) coming from different populations.

Nullisomy, the condition of having both members of a pair of chromosomes missing. Such a condition would almost always be lethal, as none of the genes located on the chromosome in question would be present.

O

Obesity, excess of body fat. How much fat constitutes an 'excess' has not been scientifically determined, though there are several rule-of-thumb ways of estimating it. One is the Broca index, which says that the 'ideal' weight in kilograms is the person's height in centimetres minus 100 minus a further 10 per cent for males or 15 per cent for females. Another is $W/H2$, where W is the person's weight in kilograms and H is their height in metres, the ideal ratio being 25.0 for males, 24.2 for females. (These two methods do not give the same ideal weight, as readers can discover with their own measurements.) Clinicians generally describe anyone between 10 and 19 per cent over the ideal weight as *overweight*, and anyone more than 20 per cent over as *obese*.

Fat is laid down when the food intake is surplus to the body's energy requirements; in practice, this is almost always because of eating too much rather than exercising too little. Obesity presents a >nature versus nurture problem almost as great as that of the inheritance of >IQ. Fatness does indeed run in families, but is it in the genes or the lifestyle? Most fat people have at least one fat parent, but the fact that this may equally often be the father or the mother points towards a genetic element. It has been observed that people with a skeletal framework more circular in cross-section tend to be more obese than those with a flattened-front-to-back frame, the frame shape being presumed to be predominantly inherited. It has also been found that, while mice of most strains become obese when fed on a high-fat diet, there are two strains that do not. None the less the outcome of studies on twins and on adopted children (*see* twin studies; adoption) has shown that the heritability of obesity is rather low.

In a tiny minority of cases, obesity is due to a specific physiological defect, possibly genetic in origin. This is the case with the Laurence–

Moon–Biedl syndrome in humans, where there is a defect in the hypothalamus (a part of the brain concerned with regulating body functions), and with three other genetic syndromes. There are three different >mutations in laboratory mice that cause obesity, each via a different >inborn error of metabolism, and it is likely that each of these also exists in humans.

Modern medical opinion is that it is not sensible to treat patients for obesity if the distress caused to them by dieting is disproportionate compared to the risk to their health of staying fat; this is particularly the case with late-onset (40-ish) obesity, in which the extra risk of, say, heart disease is in fact small.

Objectivity. *See* quantification.

Ochre codon, one of the three >nonsense codons in DNA that code for 'stop' and therefore signal the termination of the protein being synthesized. The ochre codon is UAA.

Oenothera, the evening primrose, a plant much studied in genetics. It has an interesting arrangement of chromosomes, which are joined end to end in one large ring rather than lying in pairs. It is no relation of the British wild primrose (*Primula vulgaris*).

Ohno's rule, a theory that, because the X chromosome has intrinsic properties of gene regulation, certain genes are always situated on it, whatever the species. An interesting example of this is the fact that the mutation that causes haemophilia is found in dogs as well as in humans, and is on the X chromosome in dogs as it is in humans. It is called after the American geneticist Susumu Ohno.

Okazaki fragments, short sequences of DNA that are formed on the 'backward-facing' strand during >replication and are joined up by ligation. They are called after their discoverer, the Japanese geneticist Reiji Okazaki. *See* diagram at replication.

Oligogene, a gene with a major effect on the phenotype. Many are genes that control some part of the developmental process, and are sometimes referred to as 'switch genes'.

Oligonucleotide, a sequence of >nucleotides (as in DNA or RNA) but containing only a few nucleotides. Such a sequence is more usually synthetic than naturally occuring.

Oncogene, a gene that can cause a cell to become a tumour cell. Originally identified in >retroviruses, oncogenes have now been recognized as part of the normal genotype of most cells. They are probably concerned with the control of cell growth and division. The oncogenes found in cells are known as proto-oncogenes, to indicate the fact that they are only potentially, not actually, oncogenic (i.e. cancer-causing). They become cancerous when something interferes with their normal regulation system. Viral oncogenes are proto-oncogenes which during the evolution of the virus have been picked up from the genomes of host cells and have become changed so that can cause cancer. *See* cancer and pp. 33–4.

Ontogeny (from the Greek *ont*, part of the verb 'to be'), the sequence of development of the individual organism from fertilization to maturity. *See* Haeckel's law.

Oogenesis, the process by which female organisms make an ovum, or egg. In humans, the eggs are usually matured in the ovary one at a time (if more than one egg is matured simultaneously the result may be a multiple birth). A primary oocyte (immature egg) goes through the first division of >meiosis to produce a secondary oocyte with a non-functional >polar body attached to it; the second meiotic division produces the mature ovum with a second polar body. When a female human is born, all the oocytes that she will ever have are already present; this may be as many 500,000, vastly more than will be needed in 30 years of reproductive life. The details of oogenesis differ from one species to another, and are quite different in plants. Compare >spermatogenesis.

Opal codon, one of the three >nonsense codons in DNA that code for 'stop' and therefore signal the termination of the protein being synthesized. The opal codon is UGA.

Operator, a part of the >operon. The operator region, immediately upstream of the structural genes, is where transcription of the rest of the operon is blocked or not, according to whether the repressor is bound there or not.

Operon, a group of genes all coding for parts of a metabolic pathway (a group of enzymes that work one after another to control a metabolic process within a cell), plus their associated regulatory apparatus. The parts of the operon lie adjacent to one another in consecutive order, with only short non-coding sequences interspersed. Credit for the discovery of the operon goes to the French molecular biologists Jacques Monod and François Jacob, who reported it in 1961 (*see lac* operon). The parts of the operon are: *(1)* a gene for the repressor molecule (a protein); *(2)* a promoter region; *(3)* an operator region; and *(4)* the structural genes, in a row. RNA polymerase that will transcribe the operon binds to the promoter region; if the operon is inactive, the repressor molecule is bound to the operator region immediately adjacent, and in that case, the RNA polymerase is unable to proceed so no transcription takes place.

In the presence of the inducer molecule (very often the substrate of the enzymes in question), the repressor combines with the inducer and dissociates itself from the operator region, thus allowing the RNA polymerase to proceed and transcribe the structural genes. This goes on until

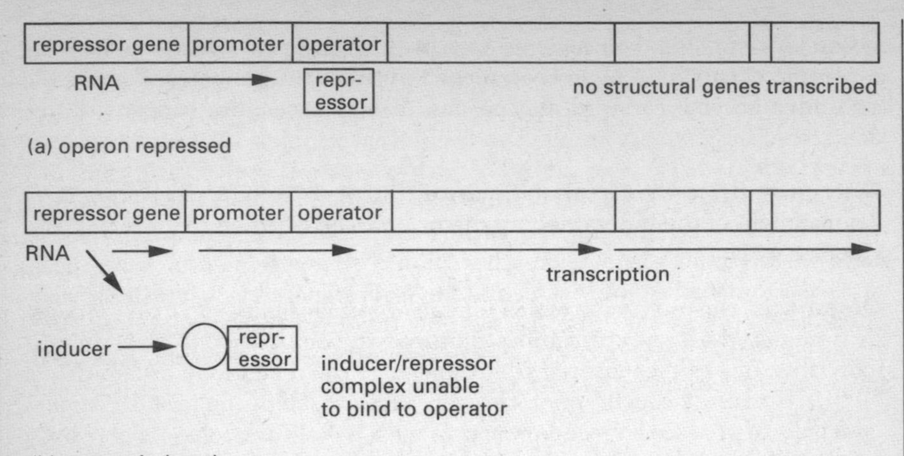

there is no more of the inducer in the cell, so the repressor binds on to the operator once more, blocking transcription.

This is the process in operons that are *negatively inducible*. In *positively inducible* operons, there is no bound repressor, but the same function is carried out by a protein co-activator (a small molecule, analogous to the inducer) and an apo-activator (a protein, analogous to the repressor), which combine to allow RNA polymerase to pass the operator region. In repressible operons, transcription goes on continuously until the process is shut off by the binding to the operator region of a co-repressor (small molecule) combined with an apo-repressor (protein).

Operons occur particularly where the enzymes are involved in only one metabolic process, so that they are needed at the same time and in the same quantities; in other cases, genes for related enzymes may be scattered at separate loci. Numerous operons have been found in prokaryotes, and are assumed also to occur in eukaryotes though the evidence for this is lacking.

Organelle, any structure within a cell that has a membrane round it and has a particular function in cell metabolism. *See* eukaryote.

Origin of life, an event that may have taken place a number of times. There is a statistical argument in favour of life existing on other planets (*see* exobiology). Life may have started on this planet more than once, and come to nothing, but it is overwhelmingly likely that all life as existing today had a single origin, the evidence being that the genetic code is identical in all life forms. Darwin wrote in 1871: 'If (and oh what a big if) we could conceive in some warm little pond, with all sorts of ammonia and phosphoric salts, light, heat, electricity, etc., present, that a protein compound was chemically formed . . .' This was taken literally more than 80 years later by the American physical chemist Harold Urey; in 1953, he and his research student found that a gas consisting of water vapour, methane, ammonia and hydrogen will, if exposed to electric sparks as generated by a thunderstorm, produce various organic molecules including amino acids. Later work has produced even more encouraging results, though nothing approaching a molecule as complex as a protein or a nucleic acid (i.e. DNA or RNA). An alternative theory has it that life originated elsewhere and was brought to Earth on a meteorite. However, this does not solve the problem of how that life form was originated; it merely shifts the scene to another planet. If experimental work does succeed in producing life – which is far from impossible – it will be very interesting (to say the least) to see how it differs from the system that already exists. The first origin of life was an exception to the rule that spontaneous generation cannot occur, and so will the second one be – if it happens.

Ortet, the original organism from which a >clone is derived.

Outbreeding, a mating system in which parents are not closely related to each other; technically called >exogamy. This is the system that obtains in nearly all natural populations.

Overdominant, a dominant allele which, in the heterozygous state, causes the trait to be more strongly expressed than it does in the homozygous state. For example, in a plant with allele *C* for curly leaves and allele *c* for flat leaves, the heterozygote *Cc* would have curlier leaves than the homozygote *CC*. The phenomenon is related to hybrid vigour.

Overlapping genes, genes that have coding sequences in common, overlapping each other. The two sequences are translated with a different >reading frame. This system is highly economical of space, and occurs mainly in bacteriophages which have a very small genome. It does not occur at all in eukaryotes.

Oviparous, bringing forth young in eggs. Generally this is a primitive evolutionary feature, shared by most invertebrates, fishes, reptiles and amphibians and all birds. The monotremes (an anomalous group containing the duck-billed platypus and the spiny anteater) are the only egg-laying mammals. An immature form of the adult may hatch directly from the egg, or there may be a larval stage. *See also* ovoviviparous; viviparous.

Ovoviviparous, bringing forth live young that have been hatched from eggs inside the mother's body. Some invertebrates, fish, reptiles and amphibians are ovoviviparous. *See also* oviparous; viviparous.

Ovum, the >gamete produced by the female. The mature human ovum is about 0.15 mm in diameter. *See also* oogenesis.

P

p, symbol used to designate the short arm of any chromosome. Thus when a gene's locus is given as 11p, this means that it is on the short arm of chromosome 11. The long arm is called >q.

P_1, P_2, P_3, etc., the parental generations in a cross. The parents in the first cross are P_1, and their offspring are the >F_1 generation. P_1's parents are P_2, P_2's parents are P_3 and so on.

P450, a gene superfamily, consisting of large groups of genes that code for proteins (cytochromes) which metabolize drugs or other foreign chemicals. The P450 cytochromes are coded for by 20 different families of genes, ten of which are the same in all mammals. Any one mammalian species may have between 60 and 200 individual P450 genes, each making a unique cytochrome. The drug or chemical acts in many cases as the >inducer for the relevant P450 gene. In humans, polymorphisms of P450 genes are responsible for many of the examples of patients' variable responses to clinical drugs.

Pachytene, one of the stages in >meiosis (one of the two processes of cell division). *See* diagram at meiosis.

Paedomorphosis (from the Greek for 'child-shaped'), the retention in the adult of some feature that was appropriate to the juvenile of the ancestral stock. The human skull shape has been described as an example of paedomorphosis, in that the rounded cranium and small vertical face are features of an ape child. *See also* neoteny.

Palaeontology, the study of evolution as shown in the succession of fossils. It predates the theory of evolution by natural selection, though it did contribute to formulation (*see* succession, principle of). However, Cuvier, the most notable early palaeontologist, was a creationist (*see* p. 14). Modern palaeontology interprets fossil evidence in the light of genetical and ecological principles.

Palindrome, a sequence of DNA that reads the same in either direction, with a central point of symmetry. Such sequences are often the target sites for regulatory molecules or for cleavage by >restriction endonucleases.

Pangenesis, a pre-Mendelian theory of heredity. It proposes that each cell or tissue of the body produces a minute copy of itself (called a 'gemmule') which is sent to the ovary or testes to be assembled into a gamete and transmitted to the offspring (this is quite like Hippocrates' idea of a genetical fluid; *see* p. 8). Darwin adopted the idea of pangenesis, in the absence of a more plausible mechanism of inheritance. It was, however, soon disproved by Galton, who carried out blood transfusions between black and white, and observed that no mixing of gemmules took place: the theory predicted that the rabbits' offspring should be speckled or dappled, but they were not. Pangenesis was essentially a refinement of >preformation, in that the gamete was supposed to be a conglomerate of minute particles from each of the tissues present in the adult body.

Panmixis, unrestricted >random mating within a closed population, there being no contribution to the gene pool by migration from outside.

Papovaviruses, a group of viruses whose genome is a ring of double-stranded DNA. One of them is SV40, a virus originally identified in green monkeys and now extensively used in virus research. The whole sequence of its genome, 5224 base-pairs long, has been decoded.

Paracentric. *See* inversion.

Parallelism, evolutionary change in which two or more lineages derived from a common stock retain similar features even though they are developing in isolation from one another. An example is the two groups of monkeys – the Old World monkeys (Cercopithecoidea) and the New World monkeys (Ceboidea) – which have been separate for 35 million years but are still very similar in general anatomy. *See also* adaptive radiation; convergence.

Paramecium, a single-celled eukaryote (one of the Protozoa), very common in fresh water. It is interesting genetically because it has two nuclei in its cell. The larger macronucleus is used when *Paramecium* reproduces by simple division, and the smaller micronucleus is only used during conjugation.

Paramutation, an allele that appears to have mutated but has in fact no change in its DNA sequence. The change in phenotype that it seems to cause is due to activity by the other allele at the same locus. The condition is extremely rare, but has been much studied in maize (*Zea mays*).

Parasexual reproduction, a cycle of reproduction that occurs in some fungi as an alternative to their normal cycle via spores. The process involves the fusion of two haploid nuclei, a mitosis division with crossing-over (not a meiosis division), and a return to the haploid state.

Parthenogenesis (from the Greek *parthenos*, 'young woman' or, sometimes, 'virgin', and *genesis*, 'creation' or 'birth'), the process of reproduction in which an organism develops from an egg without its being fertilized. Parthenogenesis can be artificially induced, by physically or chemically interfering with an ovum to stimulate it into division; the resulting individual is haploid. It also occurs frequently in nature. One example is found in all the social insects of the order Hymenoptera, such as wasps, bees and ants. Here females are diploid whereas males are haploid; any unfertilized egg develops as a male, and a fertilized one develops as a female. The colony of insects is able to adjust its own sex ratio by choosing how many eggs to fertilize. Aphids have an optional phase of parthenogenesis that they use when they want a rapid increase in population at the beginning of summer: they produce large numbers of diploid eggs that develop without fertilization (these eggs are diploid as the result of a special modification of meiosis in which only the first meiotic division is completed). Some species of fish, amphibians and lizards are also capable of parthenogenesis. *See* sex determination.

Partial dominant. *See* semi-dominant.

Passive immunity, immunity to a disease or toxin that is not produced by the individual's own immune response. Passive immunity can be given to a patient by injecting a substantial quantity of >antiserum. This contains antibody to the antigen in question, so that immunity starts immediately without waiting for the patient's own immune system to produce antibody. One example is the prevention of tetanus by injecting antiserum after an injury, and another is the use of anti-Rhesus antibody to prevent haemolytic disease of the newborn. Passive immunity also exists in nature, as the mother passes on some immunity to her foetus through the placenta and later to the newborn baby in colostrum and breast milk.

Pasteur, Louis (1822–95), French chemist. Only a moderate student at school, Pasteur intended to be a teacher until he accidently discovered chemistry at college. His first research, in 1848, concerned the two forms of tartaric acid, which he discovered were dissymmetrical (i.e mirror images); this was the beginning of stereochemistry. Next he became interested in fermentation (a subject of great practical importance in wine-producing France), and in 1857 he proved that fermentation of a sugar solution or putrefaction of a broth both depend on the presence of micro-organisms (yeast or bacteria). This was the final disproof of >spontaneous generation, and is Pasteur's most important contribution as far as genetics is concerned.

This work led to his invention of what came to be called 'pasteurization', heating fluids for 30 minutes at about 65°C (150°F) to sterilize them. It also led to his proposition that all diseases and infections are caused by micro-organisms. This was put to immediate practical use in his investigation into the diseases of silkworms that were ruining France's silk industry; Pasteur quickly showed that bacteria were responsible. He then extended Jenner's work on >vaccination (a word coined by Pasteur), producing in 1880 a very successful vaccine against anthrax (then a common disease), and later one against rabies. Although Pasteur's discoveries were of vast economic importance, he refused monetary reward and lived simply.

Paternity, the relationship between a male and the offspring that he has sired. Humans are not the only species who hold paternity to be important: in lions and in langurs (arboreal monkeys of southern Asia), it has been observed that if a new male takes over a harem he will kill all the infants sired by his predecessor. Paternity in the sense of a child-caring role for the father is widespread among birds: 90 per cent of species are monogamous and share the care of the offspring. For a discussion of human attitudes, *see* pp. 44–5. The technique of establishing whether someone is or is not the father of a particular child has greatly improved with >genetic fingerprinting.

Pauling, Linus Carl (1901–), American chemist; has taught at Stanford University, California since 1969; director of the Linus Pauling Institute of Science and Medicine from 1973. Pauling's work covers an enormously wide range, but from the point of view of genetics, the most important areas of his research are: *(1)* his discovery of the alpha-helix structure of some proteins (which was a clue to Crick and Watson in their search for the structure of DNA); and *(2)* his suggestion that the deficient haemoglobin in sickle-cell anaemia is due to a change in the amino acid sequence. Pauling believes that large doses of vitamin C are effective in preventing the common cold and other illnesses. He has won two Nobel prizes, for chemistry (1954) and, as a vigorous proponent of nuclear disarmament, for peace (1962).

Pea (*Pisum sativum*), the ordinary culinary pea, which was the subject of Mendel's historic experiments (*see* pp. 18–21). Mendel studied seven qualitative traits of pea plants (listed here with dominant variant first): flower colour (purple or white), seed shape (round or wrinkled), cotyledon colour (yellow or green), pod shape (inflated or constricted), pod colour (green or yellow), flower position (axial or terminal) and stem length (long or short). From looking at the way these behaved in crosses, he discovered the basic machinery of inheritance. The pea has seven pairs of chromosomes, and the fact that the gene for each of Mendel's seven studied traits was on a different chromosome was either a 1-in-166 chance or was due to the fact that Mendel already knew what he was looking for.

Pearson, Karl (1857–1936), British statistician. Trained in law, Pearson soon turned to mathematics, and became professor of mathematics at University College, London during Galton's time there. His work in the theory of probability and distributions is still the foundation of statistics. He defined the standard deviation of a distribution, and devised the >chi-square test.

Pedigree, a diagram representing the genetic relationships between individuals. Usually drawn up with reference to one (or a few) specific traits, the pedigree uses conventional symbols to show the sex of the individuals (usually squares for males, circles for females); a diagonal line through their symbol shows that the person is deceased. Whether or not the individual is affected with the trait in question is shown by having their symbol solid or open, perhaps with heterozygotes (proven or surmised) shaded.

Penetrance, the frequency with which a dominant allele is expressed in the phenotype of the individual carrying it. A completely penetrant allele is expressed in every case, and this is normally the position with dominant alleles having major effects. An allele is said to have incomplete penetrance if it is not always expressed (this will depend on the genetic background and on the environment). *See also* expressivity.

Pericentric. *See* inversion.

Persistence, the number of generations that elapse before a superseded allele is entirely replaced by a superior >mutation (*see* fixation).

Perutz, Max Ferdinand (1914–), Austrian-born British X-ray crystallographer and molecular biologist. From 1947 Perutz was director of the Medical Research Council's Molecular Biology Unit at Cambridge, where the structure of DNA was discovered. In 1953 Perutz showed that the structure of a protein could be deduced from comparisons of X-ray diffraction photographs, and using this method he solved the structure of haemoglobin, the first protein ever to have its physical structure described. His colleague John Kendrew did the same for myoglobin (the oxygen-carrying protein in muscle), and the two shared the 1962 Nobel prize for physiology or medicine. Perutz's MRC Unit (later Laboratory) employed many of the most brilliant molecular biologists in the world, including >Brenner, >Crick, >Milstein, >Sanger and >Watson. He has also been chairman of the European Molecular Biology Organization (1963–9).

Phage. *See* bacteriophage.

Phene, a 'unit' of the bodily characteristics of an individual, thought of as being equivalent to the genetic unit, the gene. Geneticists challenge the usefulness of this term, pointing out that all aspects of an organism's phenotype are actually the result of interaction between the genotype and the environment, and therefore there cannot be a pure unit of phenotype that can be mapped on to the corresponding part of the genotype. None the less some traits such as blood groups, being uninfluenced by environment, do behave as implied by the word 'phene'.

Phenetic classification, a system of classifying species by their phenotypes (visible or otherwise measurable) rather than with reference to their ancestry. *See also* phylogenetic classification.

Phenetic distance, the total amount of difference in measurable traits between any two related species. This may or may not reflect how closely they are related to a common ancestor, depending on whether there has been convergence or parallelism. *See also* cladistic distance.

Phenocopy, a phenotype caused by environmental influences that is identical or very similar to the phenotype controlled by the genotype under consideration. For example, dietary deficiency may mimic the effects of an allele that prevents the organism from digesting a particular nutrient. The characteristics of the phenocopy are not heritable.

Phenotype, the totality of the physical characteristics of an organism. The term is used in contradistinction to the >genotype, the total of the organism's genetic information. The genotype has an influence on the phenotype, modified to a greater or lesser degree by the environment. The phenotype used to be taken to mean only those characteristics that could be directly observed, but now includes biochemical characteristics such as the ability or inability to produce particular enzymes; it also includes behavioural characteristics.

Phenotypic variance, the total amount of >variance observable in a trait. It consists partly of the genetic variance and partly of a component contributed by environmental factors.

Phenylketonuria (PKU), a human disease caused by a recessive >allele. The actual mutation is a failure to produce the liver enzyme phenylalanine hydroxylase, so that phenylalanine cannot be metabolized and accumulates in the bloodstream in concentrations up to 100 times normal (some is also excreted in the urine, hence the name of the disease). These excessive concentrations cause severe damage to the brain. A baby with phenylketonuria is protected before it is born because its mother has the normal enzyme and deals with the baby's excess phenylalanine as it passes through the placenta, but at birth, the baby must immediately be put on a diet low in phenylalanine to prevent brain damage occurring. (Some phenylalanine is essential, as it is one of the amino acids necessary for protein synthesis.) All babies born in Britain are routinely screened for this disorder, which occurs in about 1 in 10,000 births. A prenatal test is now available for mothers who have already had one phenylketonuric baby.

Phenylthiocarbamide. *See* taste blindness.

Pheromones, a group of chemicals that act as messengers to alter the behaviour of other members of the same species, in a way analogous to hormones altering the activity of other tissues within the body (hence the pheromone's alternative name 'sociohormone'). Pheromones are certainly the oldest form of intra-specific signalling, and the chemical communication system has a number of advantages over the later auditory and visual ones: it is bioenergetically very economical; its signals can be transmitted in the dark and around obstacles, in air or in water; and it has by far the greatest range and duration. Pheromones have been found in every major animal group, including mammals, and though none has been identified in humans, their presence is suspected. They are particularly important in social insects, which may have up to a dozen different pheromone-producing glands, each for a different message. Sex-attractant pheromones are also very widespread, especially the ones put out by females, and can often operate over a distance of several kilometres. Territory and trail markers are common too. Odours to indicate dominance status are found in mammals. There has to be a simultaneous evolution of both the production and the reception apparatus within the species for a pheromone system to work. *See* allomones.

Philadelphia chromosome, an abnormal chromosome found in the cancerous white blood cells of those suffering from chronic myeloid>leukaemia. In these cells, there is a translocation of part of chromosome 22 to chromosome 9; all the rest of the cells are normal. It seems that what happens is a genetic accident within a single white blood cell; that cell is transformed into a cancer cell which then proliferates out of control. *See* pp. 33–4; and oncogene.

Photoperiodism, biological activity that is influenced by changing day length. Many plants, especially flowering ones, are photoperiodic, being generally sensitive to the length of the dark period rather than the light (the two can be experimentally manipulated to be independent). Bird migrations are mostly also photoperiodic. The system probably evolved as a more reliable way of timing hazardous operations such as flowering and migrating than by temperature: a sudden spell of unseasonable weather could otherwise deceive individuals into acting at the wrong time.

Photoreactivation. *See* DNA repair.

Phyletic evolution, evolutionary change in which one species transforms over time into another species, without any branching of the ancestral tree; there are never two different species at the same time. An example is the emergence of *Homo sapiens* from the earlier species *Homo erectus*. *See also* cladogenic evolution.

Phyletic gradualism. *See* gradualism.

Phylogenetic classification >a system of classifying species and higher taxa according to their ancestral relationships (in contrast to >phenetic classification, i.e. by observable characteristics). It has been noticed that classifications made by peoples without knowledge of evolution, e.g. Australian Aborigines, are very much closer to the phylogenetic rather than the phenetic system, contrary to expectation.

Phylogeny, the relationship between a species and its evolutionary predecessors. *See* Haeckel's law: 'Ontogeny recapitulates phylogeny.'

Pigmentation, any colouring in the tissues, but particularly the external tissues such as skin and hair. The molecule that causes the black or brown colour in skin and hair is >melanin, and the concentration of this in the cells determines the actual colour seen.

Although it is obvious that populations in the tropics (or whose ancestors came from the tropics) have strong pigmentation, the evolutionary reason for this is a matter for guesswork. It is said that the original human colour was black (though there is absolutely no evidence) and that this colour is advantageous because of the protection it gives against >ultraviolet light, which causes skin cancer; but was the risk of skin cancer ever large enough to be a selective agent? Or it has been suggested that vitamin D synthesis is involved; the vitamin is synthesized in the skin, and without the shield of pigment, an unhealthy excess would be produced under the tropical sun. Either or both or none of these may have been factors. The genes responsible for skin pigmentation have not been identified, but analysis of the amount of variation suggests that alleles at only three or four loci are involved.

As for hair pigmentation, there is no theory as to why the vast majority of humans, living in environments ranging from the tropics to the Arctic circle, should have black hair, while blonde, brown and red hair colours have evolved in Europe.

Plaque, a small patch within a culture of bacteria (a 'bacterial lawn') where there are no bacteria growing. This is because the cells are being lysed (*see* lysis; bacteriophage) or because they are sensitive to an antibiotic being tested.

Plasmid, an extrachromosomal genetic element, consisting of a small ring of double-stranded DNA. Plasmids replicate themselves independently of the chromosomes, and segregate either regularly or irregularly during cell division. They occur in widely varying numbers, from one per cell to a hundred or more. Most commonly found in bacteria, they often carry genes for resistance to antibiotics (*see* R plasmid). Bacterial plasmids are used in >gene cloning. Plasmids are uncommon in eukaryotes, but do occur in the mitochondria of yeast and some plants.

Plasmodium falciparum, the protozoan parasite that causes >malaria.

Pleiotropy, the condition in which a single mutant allele causes a number of different effects. These effects may be very diverse in their end result, but are presumably due to a 'pedigree of causes' in which the primary effect of the mutation, especially if expressed early in embryonic life, leads to a chain of adverse consequences. Many of the severe human disorders are examples of pleiotropy, e.g. Marfan's syndrome, in which the primary defect in collagen synthesis leads not only to skeletal deformities but to heart and eye defects.

Ploidy, the number of multiples of the basic set of chromosomes that are present in a cell or organism. The basic set, with one copy of each chromosome, is called >haploid and is found in gametes and some simple plants. The set with two copies of each chromosome is called >diploid and is found in the cells of almost all organisms. Many species of plants tend to >polyploidy, when there are more than two times the haploid number. The word *-ploid* can be attached to a number; for example, 16-ploid means having 16 times the haploid number of chromosomes. *See also* aneuploidy.

Plus and minus strands, designations for the strands of DNA and RNA in their role in transcription and translation. The coding strand of DNA is a 'minus', and the messenger RNA taken from it is a 'plus' strand.

Point mutation, a mutation involving the substitution of just one base-pair for another. The term is also used to mean a mutation giving rise to a new allele, in contradistinction to a genetic event involving a change in chromosomal structure.

Poisson distribution, a distribution that represents the probability of occurrence of independent events. The frequency of cross-overs is fairly close to a Poisson distribution, except that one cross-over interferes somewhat with the occurrence of another, so they are not truly independent events. The Poisson distribution was described by the French mathematician Siméon Poisson (1781–1840), professor of pure mathematics at the École Polytechnique in Paris.

Polar body, a non-functional by-product of meiosis in vertebrate females. When the primary oocyte (immature egg) goes through the first meiotic division, the chromosomes are evenly divided into the two new cells but the cytoplasm is not: one cell is large, and will become the secondary oocyte, but the other has virtually no cytoplasm and becomes a mere bump on the surface of the oocyte. The secondary oocyte then goes through the second meiotic division and again produces one cell (now the ovum) with all the cytoplasm and a polar body which joins the first on the cell wall of the ovum. In some species the first polar body goes through a second meiotic division itself so that in the end there are three polar bodies. Each polar body has a haploid set of chromosomes, and is genetically different from the ovum and from the other polar bodies.

Poly-A tail, the sequence of adenine (A) bases at the >3′ end of a molecule of >messenger RNA, usually consisting of 100–200 As. Its function is thought to be stabilization of the RNA molecule. *See* terminal transferase.

Polycistronic RNA, a molecule of messenger RNA that includes the transcription product of more than one gene. These are produced when an operon is being transcribed. In other cases, they occur where structural genes are close to each other with little intervening matter, though this kind of polycistronic RNA may be cut into separate messages for translation (*see* protein synthesis).

Polycystic kidney disease, a disease in which the kidney develops multiple cysts (abnormal, fluid-filled swellings), with probable total kidney failure in middle age. Polycystic kidney disease is inherited as a dominant, and is quite common in European populations, affecting 1 per 1250 babies; the majority of patients who need treatment by kidney dialysis have this disease. The gene has been located to chromosome 16, where it is close to the gene for alpha-globulin (one of the components of haemoglobin). This genetic marker can now be used (in some families) for predicting whether a foetus has the polycystic gene.

Polyembryony, the production of more than one embryo from a single fertilized egg. The commonest example is >monozygotic (identical) twins, but another is provided by the nine-banded armadillo which has litters of four genetically identical young all derived from a single fertilized egg.

Polygenes, a group of genes, not separately identifiable, that are involved in the control of a >continuously varying trait. An individual polygene is a Mendelian allele like any other. Each polygene may have a simple additive effect on the phenotype, or there may be dominance or other interaction between the polygenes; there will also be interaction with the environment. The number of polygenes involved in any one trait can be as small as two, or as large as 100 or so, though in most cases it is probably under ten.

Polymerase, any of the enzymes that construct a polymeric molecule of nucleic acid – i.e. DNA or RNA – by joining nucleotides together. DNA polymerase makes a new strand by using a single strand of DNA as a template (*see* replication). There are different RNA polymerases for building messenger, ribosomal and transfer RNA. *See also* reverse transcriptase.

Polymerase chain reaction (PCR), a technique that allows a DNA fragment to be analysed without having to be cloned first. Two primers have to be prepared, which are oligonucleotides (very short synthetic DNA sequences, about 30 bases long) that are copies of sequences known to occur in the region of the DNA to be studied. The original DNA in the genome is subjected to melting to reduce it to single-strand state, in the presence of large quantities of the two primers; the primers find their complementary sequences on the genomic DNA and hybridize to them. A polymerase called *Taq*, which can work at high temperatures, is added; it synthesizes a strand of DNA, starting from the end of the annealed primer and using the genomic DNA as a template. The resulting double-stranded DNA is melted again, and the process repeated. The DNA in the region between the two primers is doubled each time the cycle is repeated: after one cycle, there is twice the quantity; after two cycles, four times; three cycles, eight times and so on. As each cycle takes only about five minutes, it is possible to produce large quantities of DNA within a few hours. This can then be analysed using >restriction enzymes. A DNA sample from only a single cell or a few cells is enough to carry out PCR amplification, and results can be obtained within 24 hours.

Polymorphism, 'The occurrence together in the same locality of two or more discontinuous forms of a species in such proportions that the rarest of them cannot be maintained merely by recurrent mutation.' This definition by the British geneticist E. B. Ford is one of the classic texts of genetics. By 'discontinuous forms' Ford did not mean types that would be visually distinct, but rather types having different genotypes at a particular locus. For example, the >ABO blood group system is a polymorphism, and people with blood groups A, B, AB or O are the discontinuous forms. In higher eukaryotes, about 10 per cent of loci are polymorphic. This may be a stable situation in a population, maintained because the heterozygote has superior fitness compared to the two homozygotes (*see* heterozygous advantage). More usually the polymorphism will be transient, the population being part way along the course of having a new mutation replace the old allele. Some geneticists believe that many polymorphisms are maintained merely at random, the mutant allele being neither better nor worse than the original so that their relative frequencies fluctuate by chance (*see* genetic drift).

Polypeptide, a molecule that is a chain of up to about 50 amino acids (above that it would be referred to as a >protein). It may be an independent molecule in its own right (e.g. a >hormone) or it may form part of a protein.

Polypheny, the occurrence in a population of two or more distinct phenotypes not due to underlying genetic differences. The phenotypic variability is due to environmental effects, particularly ones happening during the individuals' development (*see* epigenetic landscape). An example is the garden hydrangea, whose flowers are either pink or blue according to whether the soil is alkaline or acid.

Polyphyletic group, a group of taxa who have different evolutionary ancestors. Such groups occur when there has been >convergence in their evolutionary history. *See also* monophyletic group.

Polyploidy, the condition of having a chromosome complement of more than twice the haploid number. The most usual forms are tetraploidy (four times the haploid number) and higher even numbers; >triploidy is unusual. Polyploidy is common in higher plants and occurs in some lower animals, but is rare in vertebrates. Tetraploidy can be maintained as a stable condition in self-fertilizing plants; they produce diploid gametes which, on fertilization, give tetraploid offspring. The same is true for octaploids (eight times haploid) with tetraploid gametes and so on. This process has been important in the evolution of plants, being a short-cut to new species (the tetraploid plants cannot make fertile crosses with the original diploid variety). Occasionally one particular cell line in the body of a diploid organism can be polyploid, as with the >silk moth.

Polyposis coli, an inherited disease in which hundreds or even thousands of polyps (small soft growths) develop in the colon. It is caused by a dominant gene, whose mode of action is unknown. There is a high probability of the disease progressing to cancer of the colon, and in many cases removal of the colon is carried out as a preventive measure. Polyposis coli affects about one person in 10,000.

Polysome, a group of two or more ribosomes with an mRNA molecule, actively engaged in protein synthesis.

Polysomy, the condition of having more than two copies of a particular chromosome. The most common form is >trisomy.

Polyspermy, entry of more than one sperm into the ovum. Only one of the sperms joins its pronucleus to that of the ovum, while the other is absorbed into the cytoplasm. It happens only rarely, and has no genetic effect, as only one of the sperm is involved in fertilization.

Polytene chromosome, a structure consisting of a homologous pair of chromosomes fused together and with up to 1000 or so >chromatids in parallel. The DNA replicates many times to make new chromatids but without any separation; the chromatids lie alongside one another to form a wide ribbon. Polytene chromosomes are so large (up to 10 μm wide) that one can easily observe the bands of heterochromatin (condensed non-active DNA) alternating with paler interbands where genes are being transcribed (these sometimes expand into a >puff). Sites of deletions, inversions and translocations can also be seen, where one side of the polytene chromosome does not match the other. Polytene chromosomes occur in the salivary glands of *Drosophila* and other insects.

Polytypy, the existence within a species of a number of types. These are not distinct enough from each other to be separate species but are recognizably different, and may be classified as subspecies or variant forms. The types are often geographically separated. An example of polytypy is the horse *Equus caballus*, which even in its original wild forms varied between Shetland ponies and the swift horses of the Arabian peninsula.

Population, a group of individuals forming an interbreeding community. Some degree of isolation from other populations is implied.

Population cage. *See Drosophila.*

Population genetics, the study of how genes behave in populations during evolution. The mathematical theory was laid down in the 1930s, the key thinkers being Sir Ronald >Fisher, Sewall >Wright, Godfrey Hardy and W. Weinberg (*see* Hardy–Weinberg law). The key factors in determining change in >gene frequencies were found to be >mutation, >natural selection, >migration and >genetic drift. Understanding of these factors was the foundation for practical studies in ecological genetics, such as the example of how a species of moth evolved black coloration (*see* melanism) in an area heavily polluted by soot emissions from factory chimneys. But there were no new theories in the field of population genetics until it was suggested, in 1957 by the British geneticist C. H. Waddington, that evolution processes could perhaps be analysed using >game theory. *See also* ecology.

Porphyria, a human disease existing in various forms, some of them genetically determined. The sufferer produces excess amounts of haem, the compound that gives red blood cells their colour and enables them to carry oxygen; this excess is metabolized into porphyrins which can be detected in the urine. There is also a variable amount of mental disturbance, sometimes taking the form of periods of hallucinations interspersed with complete lucidity. The most famous example of a porphyria sufferer is George III (1738–1820), king of Great Britain from 1760, who may have had acute intermittent porphyria, which would explain the insanity that eventually made him unfit to rule; this form is inherited as a single dominant >allele. Another form, also a dominant, is quite common in Sweden. A third, again dominant, is found among white South Africans at a frequency of 1 in 300, all of whom can trace their descent from a single Dutch couple who went to the Cape in 1688. There is no treatment for porphyria, but sufferers can avoid things that are known to provoke attacks, such as barbiturates. Mortality can be as high as 20 per cent.

Position effect, a change in the expression of a gene as a result of its changing position within the genome (or of another gene normally adjacent to it moving away). The splitting up of an operon will cause a change of expression, but in eukaryote genes, position effects are not common, and most genes are expressed in a similar way regardless of their position.

Positive control, a form of gene regulation in which transcription cannot occur until an activator molecule binds to the DNA in order to allow RNA polymerase to proceed with transcribing the structural gene or operon. The opposite is >negative control.

Posthumous conception, one of the possibly desirable results of keeping sperm or eggs frozen for future use. Storing sperm is usually done for men about to have a vasectomy, so that they can still father children if they later want to. Posthumous conception using a dead man's sperm would be technically simple, but it is hard to imagine anyone setting up the much more complicated business of *in vitro* fertilization and surrogacy that would be involved in posthumous conception using a dead woman's egg. Those who worry about telling a child about its 'real' (i.e. genetical) parent would find either of these scenarios difficult. *See also* sperm banks.

Poxviruses, a group of rather large viruses with a genome consisting of double-stranded DNA. It includes the viruses that cause cowpox and myxomatosis; the virus that caused smallpox is now extinct in the wild but is maintained as a laboratory stock in only a handful of centres for scientific interest. Vaccination has been effective against poxviruses, as they do not evolve new unrecognizable strains.

Pre-adaptation, the situation in which a mutation gives an organism a phenotypic character which happens to be adaptive to an environmental change that occurs later. *See* adaptation.

Preformation, the idea that the human being is already fully formed at the moment of conception. Some preformationists held that the miniature human homunculus (*see* pp. 4–5) was to be found inside the sperm; others believed it was in the ovum. The latter (called 'ovists') went so far as to say that each female contained in her body the fully formed bodies of all her potential sons and daughters, and that her daughters within her body had all their offspring within their miniature bodies, and so on. If this theory were true, Eve would have held within her the actual bodies of all the humans that would ever be. Improved microscopy put the preformationists into retreat in the first part of the nineteenth century, and the researches of von Baer and others gave rise to the theory of >epigenesis, the progressive elaboration of the organism from the single-celled zygote. *See also* germ plasm theory; pangenesis.

Pre-implantation diagnosis, the inspection of an embryo fertilized *in vitro* (*see in vitro* fertilization) before it is implanted into the uterus, to check for the presence of genetic defects. It is also possible that this could be done in the case of a naturally conceived embryo, if it were removed before it had implanted, and then put back. It is necessary to remove one cell from the embryo (this does not damage the development of the embryo if it is done early enough) and to amplify the DNA using the >polymerase chain-reaction technique. The DNA can then be inspected for certain defects by using restriction enzymes to identify mutations that cause loss of restriction sites or by using allele-specific oligonucleotides. Linkage to restriction fragment length polymorphism alleles can show the probability of the embryo having a defective gene, if the family pattern of linkage is known. If these tests are to be carried out before the embryo is implanted, they must be done fast, so they must focus on defects that are suspected of being present. An alternative approach would be to bisect the embryo, creating an artificial pair of identical twins, and to implant one while keeping the other for more time-consuming tests, including the cell-culture tests required for detecting chromosome abnormalities. This is ethically controversial. In the case of X-linked conditions, pre-implantation diagnosis is used to ascertain the sex of the embryo, so that it can be destroyed if male; this too is ethically controversial, as only half the male offspring of a carrier inherit the condition.

Prenatal diagnosis, the various techniques that are used to discover whether an unborn baby has a genetic disorder. The simplest is ultrasound inspection, which can detect defects that cause substantial anatomical abnormality. >Amniocentesis is very useful in that it involves little risk to the foetus and can yield information on a wide range of genetic disorders. The drawback is that the tissue culture procedures take two or three weeks, with the cytological tests for chromosome abnormalities (including such relatively common conditions as Down's syndrome) taking longest; this means that any abortion following diagnosis will necessarily be a late one. The sampling of foetal blood can be used for diagnosing blood conditions (including thalassaemia and sickle-cell anaemia), but there is a significant risk to the foetus, and it cannot be carried out until about the same time as amniocentesis (12 weeks plus). A newer technique is >chorionic villus sampling. Samples of the foetus's >DNA obtained in this way can be used for direct diagnosis (*see* polymerase chain reaction) of a number of genetic disorders that can be identified because they involve the loss of a cutting site for a restriction enzyme, or for indirect prediction that the foetus carries a mutant gene because it is known to be linked to a marker (*see* restriction fragment length polymorphisms).

Pribnow box, a sequence found in prokaryote genes. Part of the >promoter region, it occurs precisely 10 base-pairs upstream from the point at which RNA transcription is to start. Eukaryotes have a >TATA box. It is called after its discoverer, the American biochemist David Pribnow.

Primary sexual characters, the reproductive organs, both internal and external. In some animals, including humans, there are also >secondary sexual characters.

Primates, the group of mammals comprising humans, apes, monkeys and prosimians (lemurs, tarsiers, etc.). The term was devised by Linnaeus, who certainly intended it to have a sense of 'first and highest', witness the fact that he also coined the words 'secundates' for mammals other than primates and 'tertiates' for all other animals.

Primitive streak, one of the layers of cells in the very early stages of >embryogenesis; from it the skeleton, muscles, heart and blood system and several other tissues are derived. The time of its appearance (at about 14 days after fertilization) has become, in the Human Fertilization and Embryology Act 1990 in Britain, the limit on the time during which embryos can be experimented upon.

Primrose (*Primula vulgaris*), a plant studied by Darwin for its mechanism of >self-sterility. The evening primrose is >*Oenothera*.

Probability, a measure of how confident one can be that a certain event will occur. It ranges from $P = 1$ (absolute certainty that it will occur) to $P = 0$ (absolute impossibility). An estimate of probability is often used to indicate the significance of statistical results. For example, if two samples have different mean values for some trait, it may be that the probability of their having come from the same population, with the difference in means arising only by chance, is $P = 0.001$, i.e. one in 1000.

Probe, a synthetic DNA or RNA sequence used to identify fragments in a sample (*see* Southern blotting). Synthetic DNA sequences can be made up to about 20 base-pairs long; this is enough to mimic a gene for a particular protein in some circumstances, if the sequence of the protein is known but the gene has not been isolated.

Progeny, a group of individuals all having the same two parents. The word can also be used to mean all the offspring of one individual, regardless of different mates.

Progeny testing, a technique used in the breeding of cattle and other livestock to improve the efficiency of selection. Assuming that the environments are roughly comparable, if the progeny of Cow M has on average a higher milk yield than that of Cow N, one can conclude that Cow M's genotype for milk yield is superior. This is a more accurate measure than simply testing each cow.

Progeria, a very rare inherited disease whose symptoms are an immensely speeded-up ageing process – wrinkling of the skin, loss of hair, bone shrinkage. By the age of ten or so, sufferers look as if they are 70. Death from heart disease usually occurs in the early teens. The gene responsible is recessive; its mode of action is unknown but is suspected of being concerned with DNA repair.

Prokaryote, any organism that has its genome in the form of a single DNA molecule (comparable to a chromosome), sometimes in the form of a ring. A prokaryote also has a simple cell structure, with no separate nucleus. The prokaryotes comprise bacteria and blue-green algae, and are a simpler form of life than the later >eukaryotes (*see* classification).

Promoter, the sequence of DNA immediately upstream from the coding sequence of a structural gene, where RNA polymerase binds on to begin transcription. In operons in prokaryote cells, the promoter is followed by the operator sequence (where the repressor or activator acts). In eukaryotes, there are particular promoter sequences for different genes, determining which of the various types of RNA polymerase will bind to the promoter. Promoter sequences contain marker sequences at a fixed distance from the beginning of the coding sequence: these are the Pribnow box in prokaryotes and the TATA box in eukaryotes. Promoters are either 'strong' or 'weak' according to the consensus between their sequences and those of the relevant RNA polymerase type.

Pronucleus, the nucleus of a gamete (sperm or ovum, in animals). It has one chromosome of each pair, so that at fertilization the new individual has a complete set of chromosomes in pairs.

Proofreading, the process in which DNA polymerase checks during replication that the correct base has been put in place, and removes any incorrect one.

Prophage, a >bacteriophage that has entered a bacterium and has become integrated into its host's genome but is quiescent. The bacteriophage's nucleic acid is replicated every time the host cell divides, but the genes are not transcribed to produce more phages until specifically induced by some genetic event or by radiation.

Prophase, one of the phases of the two types of cell division, >meiosis and >mitosis (*see* diagrams).

Proposita, propositus, in medical genetics, the person in whom an inherited condition was first diagnosed. The family pedigree around the proposita (female) or propositus (male) is then traced to ascertain the pattern of inheritance and to identify other persons who may have the same gene. The proposita/propositus can be a member of any of the living generations of a family. *See* genetic counselling.

Protein, a molecule that consists of amino acids strung together linearly. Typically a protein has two or more separate polypeptide chains; the sequence of amino acids (the primary structure) determines the shape that the chains form when folded together (the secondary structure) which is critical for the protein's function. Proteins can be *(a)* >enzymes, *(b)* >antibodies, *(c)* >hormones (d) structural constituents, e.g. collagen, *(e)* transport molecules, e.g. >haemoglobin, *(f)* nutritional, e.g. casein, or *(g)* contractile molecules in muscle. *See* protein synthesis.

Protein synthesis, the manufacture of proteins of all sorts from the instructions in the genes. The central dogma of molecular biology states that DNA is transcribed into RNA which is translated into protein. The DNA is situated in the chromosomes, in the nucleus of each cell.

When the protein synthesis is to start (a complicated matter in itself; *see* gene regulation), an enzyme called RNA polymerase begins to build a molecule of messenger RNA (mRNA for short) according to the code in the DNA: this works in the same way as the DNA replicating itself, i.e. each base in the DNA must be matched with the complementary base in the RNA. The mRNA goes out of the nucleus into the cell's cytoplasm; among the many things floating around in this crowded space are *(a)* molecules of transfer RNA (tRNA) and *(b)* ribosomes. The tRNA molecules consist of only three bases (i.e. a single triplet); these three bases form an anticodon, and each of these matches on to a codon in the mRNA. Thus if the mRNA reads GCA–AAG–AAA, it will attract to itself three tRNA molecules having the anticodons CGU–UUC–UUU. Each of these has the appropriate amino acid (as specified in the >genetic code) attached to its other end.

Actual assembly takes place on the ribosomes, cottage-loaf-shaped bodies made of proteins and RNA (rRNA or ribosomal RNA), which may be free-floating or embedded in a network of membranes called the endoplasmic reticulum. The linear molecule of mRNA is gripped in the groove of the ribosome, the starting point always being the codon AUG in the mRNA. The ribosome has two slots alongside each other, and a tRNA molecule with the anticodon UAC (complementary to the AUG in the mRNA) moves into the first one; then the tRNA appropriate for the next mRNA codon moves into the second slot, and the ribosome joins their amino acids together. The first tRNA molecule floats off, the second moves into the first ribosome slot, the next tRNA moves into the second slot, the amino acids are joined, and so on until the ribosome comes to a 'stop' codon in the mRNA. This is a codon that does not have a corresponding amino acid, so no tRNA fits with it and the chain comes to an end. It has already been folding itself into shape at the front end while the back was still being synthesized.

All the RNA components in this process get recycled. The tRNA molecules float off the ribosome empty, having had their amino acids built into the protein, so they go off and pick up a new amino acid. The mRNA, once used for protein synthesis, breaks up and goes back into the nucleus to be reused for a new mRNA molecule.

- For an excellent description of this, and many other things, see Larry Gonick and Mark Wheelis, *The Cartoon Guide to Genetics*, Barnes & Noble, New York, 1983.

Protoctista, one of the five kingdoms of living things (*see* classification). It includes all the single-celled >eukaryotes.

Proto-oncogene. *See* oncogene, and pp. 33–4.

Protozoa, a group of single-celled organisms that are not plants, forming part of the kingdom Protoctista (*see* classification). The amoeba, >*Paramecium* and >*Chlamydomonas* are the best-known examples.

Provirus, the genome of a virus when it is incorporated into the genome of a host eukaryotic cell but is quiescent. The viral genome is replicated during cell division but none of the viral genes is expressed. It is a characteristic of >retroviruses that they become transmitted as proviruses. *See also* prophage.

Pseudodominance, the phenomenon in which an allele that is not normally dominant behaves as if it were because the corresponding allele on the other chromosome has been removed by a deletion.

Pseudogene, a sequence of >DNA that resembles a normal gene but cannot be transcribed (*see* transcription) because of the number of nonsense mutations in it. They occur only in eukaryotes, and are often grouped together in a single region with many repetitions or nearly identical copies.

Psoriasis, a disease in which there is overproliferation of skin cells, leading to irritating rough patches; there is no direct effect on general health. Psoriasis has a genetic component, though there is no straightforward Mendelian pattern of inheritance. There is some association with the immune system, as some HLA types are more susceptible than others. Psoriasis affects 1–2 per cent of white populations.

Pteridophytes, a group of plants including the ferns, club mosses and horsetails that reproduce without seeds. They have alternation of generations, as seed-bearing plants do.

Puff, part of a >polytene chromosome where one of the bands is greatly expanded. The puff consists of loops of DNA that has unwound itself in order to be transcribed. It is believed that puffs are the sites of transcription of cell-specific genes (*see* luxury gene), since the pattern of puffing is specific to the organ concerned, within the same organism. The largest puffs are known as Balbiani rings.

Punctuated evolution, evolutionary change which, over the long term, consists of periods of relative stability punctuated by bursts of rapid species formation ever short periods. This is not what was originally envisaged by Darwin and Wallace, but is perfectly consistent with their theory of evolution by natural selection.

Punnett square, a technique for predicting the outcome of a cross, invented by the British geneticist R. C. Punnett in 1906. A table is drawn up with a column or row for each of the possible parental gametes; the genotypes that will occur among the offspring are then read off from the squares. The method also shows what proportions of each genotype there will be. For an example, *see* dihybrid.

Purine nucleoside phosphorylase (PNP), an enzyme involved in the immune system. Lack of it (inherited as a recessive) causes one form of >severe combined immune deficiency.

Purines, one of the two types of nitrogenous bases that are the important constituents of DNA and RNA. The commonest (and the only ones occurring in the nucleic acids) are adenine and guanine. *See also* pyrimidines.

Pyloric stenosis, an inherited disorder in which the pylorus (the muscular ring at the lower end of the stomach) is enlarged and consequently the outlet to the small intestine is narrowed. A baby with this disorder is normal at birth but begins to vomit at about two months old; without surgical intervention to clear the enlargement the baby dies. The inheritance is polygenic, and for some unknown reason boys are about four times more likely to be affected than girls. It is a common disorder, with an overall frequency of 1 in 200–300 births.

Pyrimidines, one of the two types of nitrogenous bases that are the important constituents of DNA and RNA. The commonest (and the only ones occurring in the nucleic acids) are cytosine, thymine (in DNA) and uracil (in RNA). *See also* purines.

Q

q, symbol used to designate the long arm of a chromosome. Thus if a gene's locus is given as 22q, this means that it is on the long arm of chromosome 22. The short arm is called >p.

Quadruplets, four babies born together. The frequency with which quadruplets are born is the cube of the frequency of twins (with triplets being the square); there is no theoretical basis for why this should be so (*see* twinning rate). This frequency will no longer hold because of the high frequency of >multiple births following the use of >fertility drugs. Human quadruplets may be identical or non-identical.

The word would not be used of animals having four young at once, except perhaps in the case of nine-banded armadillos, which have four identical pups derived from a single fertilized egg.

Quantification, the application of measurement to experimental results. Sir Francis >Galton was the pioneer of quantification in genetical studies. Quantification is not equivalent to objectivity: no matter how good one's measuring apparatus or one's technique in using it, one has to decide what to measure. A brilliant hypothetical example of this was proposed by the British mathematician Alan Turing (whose theoretical work led to the development of computers while at the same time defining their ultimate limitations). Suppose that the photograph taken of an apparent dead-heat in a horse race shows that horse A has its muzzle first across the line, but horse B has a stream of saliva coming out of its mouth, which crossed the line before the muzzle of horse A. Which is the winner? Quantification in this case is measuring the presence or absence of a horse at the line at a given time, and the subjective element is deciding whether or not saliva is part of a horse. (In deciding this, one must remember that saliva is so identifiable as belonging to one individual that it has often been used as forensic evidence.)

Quantitative trait. *See* continuous variation.

Quantum evolution, evolutionary change in which new species are formed very rapidly. The circumstance in which this is likely to happen is the colonization of new habitats by small populations, in which both founder effect and genetic drift contribute to genetic change.

R

R plasmid, any plasmid carying a gene for antibiotic >resistance.

***r* selection**, the type of natural selection that operates when a population is at low density. Food supplies and other resources being plentiful, selection will favour high reproductive rates, generally achieved by large broods and early maturation, which in turn implies small body size.

This type of selection operates when a population has recently colonized a new area, or where its habitat is naturally a temporary one, as is the case with seasonal insects such as aphids. *See also K* selection; MacArthur, Robert Helmer.

Race, an informal classification of people, supposed to be based upon biological differences. It need hardly be said that the political application of this idea was and continues to be morally outrageous. The genetical perspective on the matter is this: there are differences in the frequencies of genes between one human population and another, as natural selection has adapted them to different environments. There are no cases of hard-and-fast divisions between populations on the basis of these gene-frequency differences, and attempts to classify human groups according to the available data have resulted in widely different interpretations: for example, in 1962 the geneticist Theodosiuns >Dobzhansky estimated that there were 14 identifiable groups; the year before, S. M. Garn (an American anthropologist) had said that there were 32. So much for the idea that a 'race' is a biological entity.

To get back to the matter of gene frequency, the fact is that the genes that cause the difficulty are the ones controlling a minor aspect of a person's appearance: skin colour. These genes have not been separately identified, but it is estimated that alleles at only three or four loci need be involved (*see* pigmentation). Out of a genome of 100,000 genes or so this is an infinitesimal difference; but it is true that, at these few loci, the genes for dark colour occur at a much higher frequency in populations who themselves live or whose ancestors lived in the tropics.

Colour prejudice is irrational enough, and it would be just as irrational to discriminate against Amerindian populations on the grounds that they have practically no individuals with blood group B or Rhesus allele *r*. Again these are only a very few genes out of the whole human genome – but invisible ones.

Radiation, one of the causes of raised mutation rates and of increased incidence of cancers. There is an appreciable amount of naturally occurring radiation, from cosmic rays and from minerals in rocks. Artificial radiation comes from such sources as fall-out from nuclear weapons, the by-products of nuclear energy, and medical and dental uses (e.g. >X rays). There is no safe level of radiation: the amount of genetic damage is linearly related to the radiation dose, starting with the smallest doses.

Genetic damage, in the sense of mutations arising in the gametes, is the result of the sex organs receiving radiation; in humans, males are more at risk because the testicles are external, while the ovaries of females have some protection by being internal. Males may also be more at risk because >spermatogenesis goes on all the time, whereas females store their ova already partly matured. There may also be a cultural factor, in that men have been more exposed to radiation hazards in the belief that they are less vulnerable than women (*see* leukaemia).

Radiation causes cancers (including leukaemias) by damaging the chromosomes in the cells of the body (*see* pp. 33–4). The data from the survivors of the atom bombs dropped on Japan in 1945 show that the effect of radiation is indeed linear: the greater the dose, the higher the probability of the person developing cancer. Radiation received is measured in rem (roentgen-equivalent-man). The amount that the average person receives from background radiation is about 0.1 rem per year, and on average, medical X-rays add another 0.1 rem annually. The LD50 (i.e. the dose that would kill 50 per cent of a population) is believed to be about 450 rem for humans, although some survivors from the 1945 bombs received about 500 rem.

Ramet, an organism raised as a member of a >clone but capable of independent existence. *See also* genet; ortet.

Random mating, a system of mating in which neither genotype nor phenotype influence the choice of mate. There is no systematic inbreeding or outbreeding or assortative mating. Random mating is used as the presumed mating system in many calculations in population genetics.

Reading frame, the mode of reading of a sequence of DNA or RNA, depending on the exact point of starting. For example, the sequence AGGUCGUCC could be read as AGG–UCG–UCC, but with a different reading frame, it could become GUC–GUC–C and so on. >Mutations that add or subtract one or two bases alter the reading frame (*see* frameshift mutation). *See also* overlapping genes.

Recapitulation. *See* Haeckel's law.

Recessive, an allele that does not express its characteristic unless it is present in double dose, i.e. in the >homozygote (*see* pp. 19–20). The majority of mutations are recessive, because generally they code for a defective protein which is compensated for by the normal protein from the other allele. Note that recessiveness is relative: allele a^1 can be recessive when associated with allele a^2 but dominant in relation to allele a^3 (*see* allelic series). *See also* X linkage, a special case of recessiveness.

Reciprocal cross, two crosses, in one of which the parent from strain A is the female and from strain B the male, and in the other the reverse. The resulting offspring will be identical unless there is X linkage or maternal inheritance.

Recombinant DNA, any artificially created DNA that consists of a sequence from one genetic source spliced to a sequence from another. It is now almost always DNA from two different species, but the first recombinant DNA (made in 1973 by the American biochemist Herbert Boyer) was from two different >*Escherichia coli* >plasmids. (*See* splicing.) Recombinant DNA technology is a major part of the repertoire of >genetic engineering.

Recombination, the formation of new combinations of alleles on one chromosome, following >crossing-over at >meiosis. Recombination is second only to mutation as a source of new variation for natural selection to work upon. It also enhances the effects of mutation: a newly arisen mutant allele may not be beneficial in its original setting, and may only become so when, by recombination, it finds itself acting in concert with other genes in a new and superior combination.

Suppose that there are two strains of goats: one has long hair (genotype *HH*) and long legs (*LL*), the other short hair (*hh*) and short legs (*ll*), the loci being on the same chromosome (*see* linkage). All the first-cross (*see* F_1) goats are *Hh Ll* – that is, long-haired and long-legged – with one chromosome carrying *H L* and the other *h l*. With no recombination, the F_2-cross goats would all be either *HH LL* or *Hh Ll* or *hh ll* – i.e. only the long-haired and long-legged or short-haired and short-legged types seen in the original strains. But if in a few cases during meiosis in the F_1 animals, the homologous chromosomes have exchanged sequences, the result will be a corresponding number of animals with recombined genotypes (*see* diagram at crossing-over). The recombined chromosomal groupings are *H l* and *h L*. They will be visible in the different phenotypes in crosses with the double recessive *h l*, which will result in individuals with genotypes *Hh ll* – long-legged and short-legged – and *hh Ll* – short-haired and long-legged.

Note that the original chromosomal grouping of *H L* and *h l* has also been perpetuated; recombination provides an illustration of the combination of innovation and conservatism that is often found in genetics.

Red queen hypothesis, the idea that species inhabiting the same area are in a perpetual evolutionary race with one another, because an improvement in one species' adaptedness to its environment must be at the expense of another species.

Reduction division. *See* meiosis.

Redundant code (also called 'degenerate code'), the >genetic code of DNA and RNA, described as 'redundant' because each amino acid has more than one codon to specify it.

Regeneration, the regrowth of an organ or tissue that has been lost by accident or disease. The ability to regenerate is, perhaps paradoxically, a primitive character in evolutionary history, becoming progressively lost as organisms become more complex. Many invertebrates can regenerate an entire new individual from a severed fragment; crustacea and reptiles can regenerate legs or tails. Mammals can only regenerate damaged skin, bone and nerve fibres and those only to a limited extent. They also show the related phenomenon of compensatory hypertrophy: if one kidney is destroyed, the other grows to nearly double the size to compensate (the same happens if part of the liver or pancreas is lost). Plants regenerate extremely readily, with whole new plants able to be grown from cells taken from roots, stems or leaves in most species.

Regression line, a line drawn over plotted points on a graph where there are two variables. One is the dependent variable, and the other is the independent variable; the regression line shows how much change in the dependent variable can be predicted for a given amount of change in the independent variable. It is a measure of statistical association only, and does not imply that the two variables are causally related.

Regulator gene, a gene whose product is concerned with controlling the rate of synthesis of the products of structural genes. The structural genes may be either adjacent (cis-) or somewhere else (trans-) in the genome. *See lac* operon for an example of a regulator gene as part of an operon.

Rejection, the adverse reaction of a recipient's immune system to a graft or transplant from an unrelated donor, which it recognizes as foreign tissue. The cells of the grafted or transplanted tissue are attacked, and become non-functional. The more closely the histocompatibility antigens of donor and recipient are matched, the less likely rejection is to happen. It can be controlled to some extent by immunosuppressive drugs.

Relationship, coefficient of, the proportion of genes that are shared (on average) by relatives. It depends the number of steps between them: parent–child or brother–brother is one step, and the number of genes in common is a half. For uncle–nephew or grandparent–grandchild, both two steps, it is a quarter; first cousins, three steps, one eighth and so on. The number of genes in common is halved for each step. These are average numbers with the actual number depending on the distribution of the chromosomes during meiosis (except in the case of parent-to-child, which is necessarily exactly half).

Relict population, a remainder, usually in a geographically isolated position, of a formerly widespread species.

rem, roentgen-equivalent-man, a measure of the amount of radiation absorbed by the body. There is no safe level below which genetic damage does not occur, or where there would be no risk of cancer (*see* pp. 33–4, and roentgen).

Reoviruses, a group of viruses with a double-stranded RNA genome. They have been found in human lung and intestine tissue but have not been associated with any disease; they are therefore known as 'orphan' viruses. (The name reo is from 'respiratory and enteric orphan'.)

Repair synthesis, the process in which a section of DNA that is damaged is removed by enzymes and replaced with a newly synthesized sequence. *See also* >DNA repair.

Repetitive DNA, any sequence of DNA that occurs more than once in the genome. Repetitive DNA can be moderately repetitive, when it consists of many similar copies of sequences coding for regulatory functions and for transfer and ribosomal RNA; it also includes multigene families. Highly repetitive DNA is different in that it does not seem to have a genetic function. Higher organisms have quite large amounts of highly repetitive DNA: in humans, there is one sequence that is repeated about 300,000 times (*see* Alu sequence) and guinea pigs have a six-base-pair sequence repeated several million times. *See also* selfish DNA.

Replication, the process by which a new complete molecule of DNA is made, by one single strand being used as a template for the assembly of another single strand. This is semi-conservative replication, as opposed to conservative replication, in which the whole original double-stranded molecule would act as a template for a wholly newly synthesized double-stranded one. In 1957 Meselson proved that DNA uses the former method, as Crick and Watson had earlier suggested.

In simple terms, replication begins when the two strands of DNA part in what is called a 'replication fork', and DNA polymerase begins to assemble a corresponding new strand alongside each of the original ones. DNA polymerase reads the parent strand from the 3′ end going towards the 5′ end. Because of the antiparallel arrangement of the two strands, only one is lying the right way round for the enzyme to work in this way; this is known as the 'leading strand'. On the other, the 'lagging strand', the enzyme still works from the 3′ end, but to do so, it has to jump ahead and work back in short bursts. The resulting short fragments of newly synthesized DNA (which are called >Okazaki fragments and are about 1000 bases long) are quickly ligated together to form a continuous strand. Note that in reading from the 3′ end of the parent strand, the enzyme is building the new strand's 5′ end first and working towards its 3′ end.

In eukaryotes, chromosomes are replicated by DNA synthesis taking place in both directions from each of a number of points. Once replication has been initiated at one point, two replication forks travel away from each other (at a speed of about 1 μm per minute). Each stretch between the points is known as a replicon, and these eventually join together to form the complete new chromosome. In prokaryotes, with a circular chromosome, DNA replication begins at one point and proceeds in both directions, meeting on the opposite side (or sometimes by >rolling circle replication).

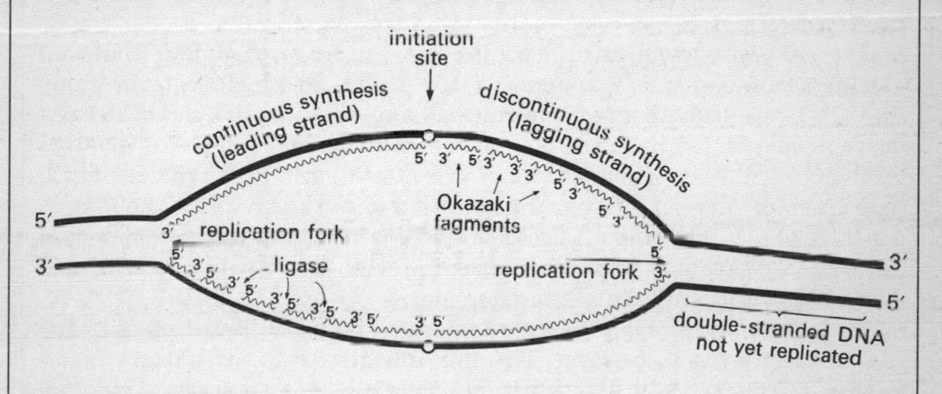

Replicon, a length of DNA between points of initiation of replication. It is not known how these points are recognized by the enzymes involved. In eukaryotes, a replicon is typically about 40,000 base-pairs long.

Repressible enzyme, an enzyme that is produced all the time, except when its gene is switched off by the binding on of the >repressor molecule to the operator. The opposite is an inducible enzyme. *See* gene regulation; operon.

Repressor, the protein that acts in an >operon to prevent transcription of the structural genes. The repressor is coded for by a sequence at the upstream end of the operon; only a very few molecules are made (perhaps only ten molecules per cell of the *lac* operon's repressor) as there is only need of enough to be sure that the operon will be turned off by one repressor molecule when required. The scarcity of repressor molecules in the cells of bacteria was the reason why it took a long time for them to be discovered and identified as proteins. The repressor is normally bound to the promoter region of the operon, but when the inducer molecule is present, it combines with the repressor so as to make the repressor dissociate from the operator, allowing transcription of the structural genes downstream to proceed.

Reproduction, the process by which a living thing creates one or more individuals similar to itself. (Viruses are not able to reproduce by themselves, they can only do so by incorporating their DNA into a host cell; they are therefore usually not considered to be living things.) The simplest method of reproduction is equal cell division, in which the parent cell divides to form two identical cells. There is no difference in the genetic content of these two cells (barring mutation). This method is used by bacteria and other very simple forms of life. A variant of this is unequal budding, which is essentially the same except that the parent cell continues its independent existence after extruding a small cell as a bud; this happens in yeasts. Again there is no genetic difference between the cells involved.

The earliest system to incorporate any exchange of genetic material during reproduction is >conjugation, the process in which bacteria transmit DNA to one another on >plasmids. Sexual reproduction can be defined as reproduction by the fusion of a male and a female >gamete. It involves exchange and reassortment of genetic material, and thus provides much more genetic diversity for natural selection to work upon. It is the system used by plants and animals (note that it does not imply an act of sexual intercourse: sexual reproduction involving wind-borne pollen fertilizing a distant flower is still sexual reproduction).

'Reproduction is by far the most important biological function that any living species has to perform. The only alternative to satisfactory peformance of this function is extinction.' This quote is from the American geneticist Theodosius Dobzhansky. *See also* meiosis; recombination; vegetative propagation.

Repulsion. *See* trans configuration.

Resistance, the ability of some bacteria to resist the effects of antibiotics because of their genetic constitution. Resistance is passed from one bacterium to another during >conjugation, when a plasmid (a non-chromosomal DNA element) carrying the protective gene is transferred. Usually one plasmid codes for resistance to a particular antibiotic, and to be resistant to more than one, a bacterium must have a number of R plasmids (as resistance-conferring plasmids are called); but in some cases R plasmids code for more than one type of resistance. Plasmids can be transferred between species or even genera of bacteria. The problem from the point of view of public health is that the more extreme the measures taken to eradicate harmful bacteria with antibiotics, the stronger the pressure of selection to produce resistant strains.

Restriction endonuclease, an enzyme that cuts a DNA molecule but only at the site of a particular sequence (usually 4–8 bases) that it recognizes. Since the first one was discovered in 1968 by the Swiss molecular biologist

Werner Arber, over 350 different restriction endonucleases have been identified, and they have proved to be extremely useful tools in DNA research and >genetic engineering. Some cut the DNA straight across, leaving a 'blunt end', but others cut so as to leave a few unpaired bases on one strand (a 'sticky end'). Their function is thought to be the protection of the cell from any incoming foreign DNA (which is cut up by the enzymes). *See* restriction map; restriction fragment length polymorphisms.

Restriction fragment length polymorphisms (RFLPs), a number of different polymorphisms in humans, detectable by the use of >restriction endonucleases (specific types of enzymes). If the total DNA in an individual's genome is digested with a restriction enzyme, it ends up as a great many fragments; these can be sorted by Southern blotting, using a probe from the part of the genome that is being investigated. This probe will find a small number of fragments of defined length. For example, when the endonuclease Hind III is used to test for >sickle-cell anaemia, a probe from the beta-globin gene will find fragments of lengths 2.7, 3.5, 7.2 and 8.0 kilobases; but in most people with sickle-cell anaemia the 2.7 and 7.2 kilobase fragments are missing (because the sickle-cell mutation has knocked out the DNA sequence at two of the sites where Hind III can act). Because not all people have the same fragments, this is said to be a polymorphism. Sometimes instead of a fragment being missing it will be of an atypical length, when a mutation has inserted or deleted some DNA between the cutting sites. If these fragment-length polymorphic loci are very closely linked to a gene causing a genetic disorder, they can be used for prenatal diagnosis.

Restriction map, a diagram of the sites on a DNA molecule, in terms of where particular >restriction endonucleases have their cutting sites. By digesting the same DNA molecule with different endonucleases (all of which have different sequences that they recognize as restriction sites, i.e. the places where they cut) one can obtain a picture of the molecule as a whole. A large fragment is digested by one endonuclease and a number of smaller fragments obtained; if this process is repeated the overlapping resulting fragment can be built up into a picture of what order the fragments are in the complete molecule. The practical complication of this procedure can be judged from the fact that even so small a genetic entity as a plasmid may have about 500 restriction sites.

Retardation, the appearance of a structure in embryogenesis later than it did in ancestral forms. The opposite is >acceleration.

Retinitis pigmentosa, an inherited condition in which pigment forms in clumps in the retina at the back of the eye; these restrict the visual field and eventually cause total blindness (unless treated). There are at least four separate loci at which genes causing retinitis pigmentosa are found: at one the mutant allele is dominant, and at another it is recessive, and at a third there are two mutant alleles, one dominant and the other recessive to the normal allele. The fourth is an X-linked recessive. The overall frequency of the disorder is about 1 in 5000.

Retinoblastoma, a form of cancer that is inherited as an autosomal dominant. Children who have the retinoblastoma gene develop tumours in both eyes in early childhood unless treated (both surgery and irradiation are effective). The gene has been located to chromosome 13, and appears not to be a mutant allele but a very small deletion of chromosomal material. This deletion happens at the site of a tumour suppressor gene; when this is knocked out, the cancer can arise. Another form of retinoblastoma is inherited as an >X-linked recessive. The autosomal dominant kind occurs in about 1 in 30,000 births.

Retroviruses, a group of viruses with a single-stranded RNA genome. Their peculiarity is that they possess the enzyme >reverse transcriptase, by means of which their RNA is back-transcribed into DNA, which is then inserted into a host cell's genome. Retroviruses are used in >genetic engineering, their reverse transcriptase being a neat way to insert a gene into a host genome. Most retroviruses include a gene that is an >oncogene, i.e. it can induce cancer. Examples are Rous sarcoma virus, several viruses causing leukaemia and other cancers in mice, and HIV (human immunodeficiency virus) which is responsible for AIDS.

Reverse genetics, term used to describe the process in human genetics in which gene mapping (often with restriction fragment length polymorphisms as markers) is used to locate a gene that causes a clinical condition. The gene can then be cloned and its product identified. This is the 'reverse' of the traditional pattern in medical genetics, when the biochemical nature of the disease would be known but the location of the gene unknown. The molecules involved in cystic fibrosis and Duchenne muscular dystrophy were both found by means of reverse genetics, and it is hoped that this is a stage towards a cure for these diseases.

Reverse mutation. *See* back mutation.

Reverse transcriptase, an enzyme that can make a strand of DNA using RNA as a template. This is a contravention of the >central dogma of molecular biology, which states that the flow of information is always from the DNA outwards, not the other way round. Reverse transcriptase occurs in >retroviruses, enabling them to get their RNA genomes copied into the DNA genomes of the host bacteria. It may also be responsible for some gene copies within eukaryote genomes: a copy that lacks the introns of the original may well have been made by reverse transcription from messenger RNA (*see* Alu sequence). Reverse transcriptase is widely used in genetic engineering. It was discovered independently in 1970 by both David >Baltimore and Howard Temin.

Rhesus blood group system, a set of antigens on the red blood cells in humans (and other primates – the name comes from the Rhesus monkey, in whose blood the antigen was first isolated). The Rhesus system was discovered in 1940 by Landsteiner, the blood-group pioneer who was also the discoverer of the first known system, ABO, in 1900. There are five Rhesus antigens: C, D, E, c and e, of which D is the clinically important one. People with D are said to be 'Rhesus-positive', and those without it are 'Rhesus-negative'. According to one theory, the genetic control is at three closely linked loci, with the D locus having two alleles: *D* (dominant) and *d* (recessive): Another theory has a single locus with multiple alleles. The gene frequencies vary a great deal between populations: Rhesus-negative people comprise about 17 per cent of European populations but only about 3 per cent of people of African descent, and are extremely rare among people of Chinese descent. Rhesus incompatibility is responsible for >haemolytic disease of the newborn.

Rheumatoid arthritis, an auto-immune disease in which there is probably an element of genetic susceptibility. The joints (usually starting with fingers and toes) are inflamed and become stiff and ultimately unusable; this is due to the immune system attacking the connective tissue. Women are three times as likely to be affected as men, and some 15 per cent of women over the age of 65 have this disease. There is some relationship to the >HLA system, in that four out of five rheumatoid arthritis sufferers have HLA type DR4.

Ribosomal RNA (abbreviation: rRNA), the form of >RNA that is found in >ribosomes. In eukaryotes, there are four types (of different lengths), the genes for which occur in many copies, making up part of the repetitive DNA in the genome. Sometimes these RNA genes are also the subject of >gene amplification. The rRNA is synthesized as a continuous long molecule which has spacer RNA (coded for by >spacer DNA) between the sequences of actual rRNA. Prokaryotes have three types of rRNA, and only seven copies of the genes for each.

Ribosomes, >organelles found in large numbers in the cytoplasm of the cells of both prokaryotes and eukaryotes. They are cottage-loaf shaped, made of >ribosomal RNA and protein, and their function is to bring together the amino acids during >protein synthesis.

Rickets, a disease in which a deficiency of vitamin D leads to failure to absorb calcium and consequently to stunted or softened bones. The vitamin is found in yeast, egg yolks and fish liver, and can also be synthesized in the body under the stimulus of the ultraviolet rays in sunlight (*see* pigmentation). In its most common form, rickets is a nutritional disease, the lack of vitamin D being simply due to a poor diet. Other cases are due to a lack of exposure to sunlight. There is also an inherited form known as vitamin-D-resistant rickets, in which the inability to absorb calcium into the bones is not involved with vitamin D but rather with phosphate metabolism. This condition is an X-linked dominant, with heterozygous females showing much variation in how severely they are affected.

Ring chromosome, an abnormal chromosome that has the ends of its arms joined to each other to form a ring.

RNA (ribonucleic acid), a molecule chemically similar to DNA. The sugar that connects the bases is ribose (rather than deoxyribose in DNA), and in almost all cases, the molecule consists of a single strand. The base-pairs in RNA are guanine(G)–cytosine(C) and adenine(A)–uracil(U) (where DNA has thymine); these carry the genetic code. RNA has a central role in >protein synthesis, at several different stages. First comes heterogeneous nuclear RNA (hnRNA), which in eukaryotes is the transcriber of the DNA code; then >messenger RNA (mRNA), which carries the coded message to the >ribosomes for synthesis; next >ribosomal RNA (rRNA), a constituent of the ribosomes; and finally >transfer RNA (tRNA), which assembles the amino acids. Each is synthesized by a particular >RNA polymerase.

RNA polymerases, the various enzymes that produce the different types of RNA, by copying from a DNA strand (the >sense strand); this is the process of >transcription, and is the basis of >protein synthesis. Prokaryotes have only one RNA polymerase, but in eukaryotes there are different ones for the different RNAs: Type I synthesizes the three largest ribosomal RNAs; Type II synthesizes heterogeneous nuclear RNA and messenger RNA; Type III synthesizes transfer RNA and the smallest ribosomal RNA; and Type IV is found only in >mitochondria, which have some independent DNA.

Robertsonian translocation, a type of chromosomal translocation; also known as whole-arm fusion. The long arm of an acrocentric chromosome breaks off very close to the centromere, and becomes joined to another (non-homologous) acrocentric which has also broken close to the centromere (but on the short-arm side of it). The results are *(1)* a long metacentric chromosome and *(2)* a very short chromosome consisting of the original two short arms. The latter is often lost in subsequent divisions, but as it may well contain little genetic information there is usually no ill effect. Whole-arm fusion probably accounts for the different karyotypes of related species, when some have all acrocentrics and others have the same amount of genetic material but with some metacentrics as well as acrocentrics.

roentgen, a unit of radiation, the amount that results in 2.083×10^9 ion pairs per cm^3 of air. It is called after the German physicist W. K. Roentgen who discovered X-rays in 1895. *See also* rem.

Rolling circle replication, a mode of DNA replication occurring in circular DNA molecules (e.g. those of bacteriophages). A nick is made in one strand, and new nucleotides are added to the 3′ side of the opening, displacing the original second strand, which falls away. As this unpeels, it also gets another complementary strand by replication, and the result is a new double-stranded length of DNA much longer than the original DNA circle.

Rous sarcoma virus, a >retrovirus that causes tumours in chickens. Identified in 1910, it was the first virus to be associated with a form of cancer.

rRNA. *See* ribosomal RNA.

Rubella or German measles, a disease causing serious >congenital defects in the foetus if the mother becomes infected with it while pregnant. The risks of the foetus being damage are: infection in first four weeks of pregnancy, 50 per cent; at 5 to 8 weeks, 35 per cent; at 9 to 12 weeks, 15 per cent; thereafter negligible. Vaccination against rubella is effective.

S

S_1, S_2, S_3, etc., generations in >self-fertilizing plants, where every generation is self-fertilized.

Saccharomyces cerevisiae, common baking or brewing yeast, one of the most important organisms in human history. It reproduces by 'budding', that is, a small outgrowth appears on the cell wall and grows until it is nearly equal in size to the original cell, when a wall forms to separate off the new cell. *Saccharomyces* (which is a fungus) has also been much used in genetic research, and is one of the organisms used in commercial >gene cloning.

Salivary-gland chromosome. *See* polytene chromosome.

Saltation, a sudden evolutionary change, probably due to a major mutation. *See* punctuated evolution.

Samesense mutation, a >mutation in which there is a change in the sequence of base-pairs in the DNA but no change in the amino acid coded for (*see* genetic code). This is most likely to occur in the third base of a codon (*see* wobble).

Sanger, Frederick (1918–), British biochemist; worked at the Medical Research Council Laboratory of Molecular Biology at Cambridge from 1951 until his retirement in 1983. Sanger devised a method for analysing the amino acid sequence of a protein chain; it involved first cutting up the protein into short sequences with acids or enzymes, then labelling the 'free amino' end of each sequence. He used this technique to work out the sequence of insulin, which has 51 amino acids arranged in two chains. For that he won his first Nobel prize for chemistry in 1958. His second came in

1980, for his work on >nucleic acid sequencing: he read the whole genome of the bacteriophage Phi X 174. He has also sequenced the entire Epstein–Barr virus (*see* herpes viruses).

Satellite chromosome, a part of a chromosome that is joined to the end of the main chromosome by a very fine thread of non-condensed chromatin.

Satellite DNA, the DNA from a eukaryote that, when centrifuged, separates out as a small fraction distinct from the bulk of the DNA in the genome. The satellite fraction contains mitochondrial and highly >repetitive DNA. It is not clear why the amount of satellite DNA should vary widely between closely related species.

Schizophrenia, a severe mental illness in which the person's connection with reality is disturbed and their reason and emotions become disjointed from one another. Schizophrenia takes many forms and there is no overall agreement among psychiatrists as to what is or is not schizophrenia. There is likewise no single explanation of its cause(s), although a genetic element seems to be involved. The best evidence for this is that if one identical twin has schizophrenia, in 60 per cent of cases the other twin will develop it too, while the equivalent figure for non-identical twins is 10 per cent. About one person in 100 in Britain suffers from schizophrenia. Some recover from it and some do not, a fact which has been interpreted as showing that there are several separate diseases with similar symptoms but quite different causations.

Scion, a shoot of one plant that is grafted on to the stock of another. The two plants are genetically different and need not even be of the same species. The word's genetical use is therefore at odds with its literary use, when it means 'heir' or 'descendant'.

Secondary sexual characters, any characters shown by one sex and not the other, except the reproductive organs themselves (the >primary sexual characters). In humans, the beard is a secondary sexual character. There has been discussion as to whether the female breasts are a primary or secondary sexual character: during breastfeeding they enlarge and are part of the necessary reproductive apparatus, but at other times they could (as in other mammals) return to their former size. *See* sexual dimorphism; sexual selection.

Segregation, the separation of the two members of a pair of alleles during the formation of gametes. It is Mendel's first law (*see* pp. 19–21). He proved that for every trait the individual has inherited one allele (a word Mendel himself did not use) from each of its two parents, and that when this individual comes to make gametes (sperm or ovum), it passes its maternal and paternal alleles to the gametes equally frequently. The two alleles are passed on not only independently but also unchanged; there is no blending.

'Segregation' is also used to refer to the separation of the two homologous chromosomes in a pair during >meiosis. *See also* independent assortment (Mendel's second law).

Segregational load, the part of the >genetic load that consists of the less fit homozygotes where the heterozygote is the fittest genotype. *See* heterozygous advantage; fitness.

Selection, coefficient of, a measure of the extent to which natural selection is acting against a given genotype, i.e. reducing its relative contribution to the next generation. Denoted by s, the coefficient varies from 0 to 1: if it $s = 0$, the genotype is not selected against at all; if $s = 1$, selection is total, i.e. the genotype makes no contribution to the next generation (a lethal or sterility gene). *See also* fitness.

Selection differential, the difference between the phenotypic mean of the population and of the individuals chosen (in artificial selection) to be the parents of the next generation.

Selection limit, the point at which further change in the genetic composition of a strain (e.g. improvement in farm animals) is impossible because there is no further genetic variability available. There can be no dogs bred to be 3 inches high because there are no alleles in the gene pool of *Canis familiaris* which have that effect. Another reason for selection reaching a limit is that improvement in one characteristic is often counterbalanced by a disimprovement in another: hens may be bred to lay more eggs per unit time, but beyond a certain point this is only achieved by reducing the size of each egg, so that the total egg output ceases to improve. *See also* heritability.

Selection pressure, the intensity with which natural selection is acting upon a population to change the gene frequencies from one generation to the next.

Selective advantage, the increase in >fitness of one genotype compared to others in the same population.

Selective reduction, induced abortion of one or more foetuses out of a multiple pregnancy. The technique consists of killing the foetus with a needle, using an ultrasound scanner to guide it. It was developed for the few cases where one twin has a defect, such as spina bifida or anencephaly (failure of the brain to grow), that can be seen by ultrasound. It has also been used for eliminating some of the foetuses in a pregnancy with a large number (quads or more) implanted after *in vitro* fertilization. It carries a high risk that all the foetuses will be lost.

Self-fertilization, the process by which a female gamete is fertilized by a male gamete produced by the same individual. A primitive genetic system, it is frequently found in flowering plants and in some lower animals that are >hermaphrodites (e.g. worms, slugs and snails). In evolutionary terms it is a disadvantageous strategy, as it leads to high levels of homozygosity within a population and consequently a less flexible genetic response to environmental change; therefore various mechanisms have evolved to prevent it (*see* self-sterility).

Selfish DNA (also called 'junk DNA'), the DNA within the genome that has no apparent genetic function. In vertebrates, this can amount to as much as 90 per cent of the genome, and there has been much speculation as to why it is there. Crick and others have suggested the term 'selfish DNA', to denote that it is present in the genome for its own benefit and not for the benefit of the animal carrying it. This is analogous to the way in which a virus inserts its own genome into that of a host cell. But it is not certain that 'selfish' DNA has no function: it may have arrived in the genome for its own sake, but perhaps the recipient organism has found a way to make it useful. Selection would tend to pick up on any usefulness of selfish DNA, for otherwise it is a great burden on the cell to have so much replicating to do for no purpose.

Self-sterility, a feature of certain flowering plants, apparently evolved to avoid >self-fertilization which increases the average homozygosity in the population. The mechanism, as studied by Darwin in the common primrose, consists of these plants having two different shapes of flowers. A bee visiting the 'pin' flower type gets pollen stuck on the middle of its proboscis from the anthers half-way down the flower tube; this pollen only rubs off on a stigma if the bee visits a 'thrum' flower, which has the stigma half-way down the tube. At the same time, the placing of the anthers at the top of the tube in the thrum corresponds with the placing of the stigma in the pin. Since a single plant only has the one kind of flower, it is self-sterile. Several varieties of commercially grown fruit trees are self-sterile because they have all the same kind of flower; they need a 'pollinator' tree nearby with the opposite-shape flowers.

Semen, the fluid that contains >sperm, produced in the testes of male animals. The fluid also contains a high concentration of the sugar fructose, probably as 'fuel' for the sperm. In humans, the volume of semen per ejaculation is 2–4 ml, which contains about 200 million sperm.

Semi-conservative replication. *See* replication.

Semi-dominant, an allele that, when present in the heterozygous state, has an effect intermediate between that of the two homozygotes. A classic example (and one that Mendel did not find in his peas) is where genotype *RR* has red flowers, *rr* has white flowers and *Rr* has pink flowers: R is said to be semi-dominant, or partially dominant. *See also* semi-lethal.

Semi-lethal (also known as semi-dominant lethal), an allele that, in the heterozygous state, has a deleterious effect but is even worse, i.e. lethal, when homozygous. Very many alleles causing human disorders are loosely described as dominant when they are, in fact, semi-lethal. For example, achondroplasia (a form of dwarfism in which the person has very short limbs but is otherwise healthy) does indeed behave as a dominant in that the *aa* homozygote is normal and the *Aa* heterozygote has achondroplasia; but it can be seen from the offspring of marriages between achondroplasics that the homozygote *AA* is very severely deformed and dies not long after birth.

A semi-lethal that kills the individual before birth gives an apparently non-Mendelian ratio. For example, if the alleles are *H* and *h*, the cross *Hh* × *Hh* gives 25 per cent *hh*, 50 per cent *Hh* and 25 per cent *HH*. Since the latter die before birth, the ratio appears to be 1:2, though it is in fact the Mendelian 1:2:1 with the last group missing.

Senescence. *See* ageing.

Sense strand, the DNA strand that serves as the template for RNA and thence protein synthesis. It is also known as the 'anticoding' strand, the messenger RNA transcribed from it being the 'coding' strand. *See* genetic code.

Severe combined immune deficiency (SCID), an inherited disease in which the body has no immune response (*see* immune system).

SCID sufferers are unable to withstand any infection by viruses or bacteria; these are the 'bubble babies' who have to kept entirely enclosed from birth. The genetic cause is a recessive mutant which fails to produce the enzyme adenosine deaminase; as a result the bone marrow cannot produce any B lymphocytes. Patients can be treated by a marrow transplant if a tissue-matched donor can be found. About 1 in 25,000 babies is born with SCID, and each has only a 50 per cent chance of surviving more than six months. It is hoped that gene therapy may produce a cure for SCID.

Sewall Wright effect, the proper name for >genetic drift, named after its discoverer, the American geneticist Sewall>Wright.

Sex chromosomes, those that determine the sex of an individual (*see* sex determination). In humans, as in many other creatures, there are two sex chromosomes, X and Y. *See also* chromosome abnormalities; gamete.

Sex determination, the various mechanisms that decide whether an individual is male or female. Some plants and animals do not have separate sexes; they are >monoecious (plants) or >hermaphrodite (animals). Species that do have separate sexes are >dioecious (plants) or bisexual (animals). In animals (in the broadest sense, i.e. members of the animal kingdom) the sex is determined by the sex chromosomes. Generally, the female has two X chromosomes (XX), and the male has an X and a Y chromosome (XY). In this arrangement, it is the activity of the Y chromosome that causes maleness; the extra X in the female is not responsible for causing femaleness (and indeed probably has no effect at all – *see* Lyon hypothesis). This can be seen from the sex of humans with the abnormal sex chromosome types XXY, who are male, while those with just one X are female (*see* Klinefelter's syndrome; Turner's syndrome).

In most higher plants and in the majority of animal groups, XX is the female and XY the male. The sex with the same two sex chromosomes, is called >homogametic (from the Greek *homos*, 'the same', because all the gametes are made the same) and the sex with two different sex chromosomes is called >heterogametic (from the Greek *heteros*, 'other', because two different sorts of gametes are made). The female makes only X-bearing ova, the male makes both X-bearing and Y-bearing sperm, and is therefore the parent responsible for the sex of the offspring. In some species, a similar arrangement has females that are XX and males that are XO (i.e. with only one X and no other sex chromosome): this works the same way, with males producing either X-bearing or 'blank' sperm. In some animals, it is the female that is heterogametic; in these species, the sex chromosomes are called W and Z, so the female is WZ and the male is ZZ. In these groups, which include the butterflies and moths, all birds and most fish, amphibians and reptiles, the sex of the offspring is determined by the mother who produces W- or Z-bearing ova.

In the Hymenoptera (the social insects such as ants, wasps and bees) the females are diploid and lay haploid eggs. If the eggs are fertilized they become diploid and are therefore female, but if they are not fertilized they remain haploid and develop into males (the males produce their sperm by mitosis, not meiosis). (*See* social insects; parthenogenesis.)

In some organisms sex is determined not by chromosomes but by the environment. Some turtles do not have sex chromosomes, and develop into one sex or the other according to the temperature at which the eggs are incubated. An extreme case is the *Bonellia* marine worm: the larva settles in the mud and if it has enough space round it develops into a female, after which any other larva that arrives is turned into a male by the influence of her secretions. Maleness to this marine worm involves the loss of all organs except the reproductive one, and the male becomes a parasite inside the female's reproductive tract. Some fish, e.g. wrasses, have sex reversal, with individuals able to change from one sex to the other if the population's sex ratio becomes unbalanced. A special example of this is the Pacific coral-reef fish *Labroides* in which a male fish has a harem of individuals who are kept female by being dominated; when the male dies, the most senior female changes sex and becomes the new male.

Sex hormones, those that control the reproductive cycle in animals. The female steroid hormones are: the oestrogens, produced by the ovary and the placenta (these are used in hormone replacement therapy in women who have reached the menopause); and the progestogens, produced by the placenta during pregnancy. Males produce androgens, also steroids, of which the most important is testosterone, produced in the testes and important in controlling spermatogenesis. The androgens also have an anabolic (growth-stimulating) effect; synthetic anabolic androgens have been developed which give the extra growth (especially in muscles) without too much of the masculinizing effect. Both sexes produce gonadotrophic hormones in their pituitary glands, which act upon the ovary and the testes. One of these, follicle-stimulating hormone (FSH), stimulates the ovary to produce ripe eggs; it is used to treat some infertile women, and can cause multiple births.

Sex-limited trait, any trait that is only expressed in one sex. An example is milk yield in cows, which can obviously only be a trait of the female even though the genes for it are carried on the autosomes and are therefore present in both sexes. A more subtle example is male-pattern baldness in humans, which only affects males because it depends on the influence of the male sex hormones. *See also* X linkage.

Sex linkage. *See* X linkage.

Sex ratio, the proportion of males to females in a population. Although in most species of animals, it seems obvious that the ratio is 1:1, the position is more complicated. Males being somewhat more vulnerable during embryogenesis, many species have an excess of males at conception to allow for wastage. In humans there is excess mortality of boys during adolescence, and to ensure an approximately equal sex ratio at maturity the ratio at birth is about 106:100. Human males are also less robust in old age, so the sex ratio taken at age 70+ (in Britain) is about 74:100 (and by age 80 it is 43:100). Some species can adjust the sex ratio to suit conditions (*see* sex determination), but in general it can be shown that the sex ratio will evolve towards being equal at maturity. The following argument was put forward by Fisher: if males are rarer, each male will have a better chance of finding a mate, so it will pay parents to produce extra males; but this advantage will cease when males are no longer rare. The same argument applies to females, so the sex ratio stabilizes.

Sex reversal, alteration of an individual's sex. It occurs in nature, especially with some species of fish (*see* sex determination), and can also be artifically induced in many species by surgical and chemical intervention.

Sexual dimorphism, the existence of physical differences between the two sexes, other than the differences in the reproductive organs. Some species, e.g. most snakes, have almost no sexual dimorphism, but others, including many birds, take it to spectacular extremes. Most mammals have at least a difference in size of skeleton and musculature. In humans, sexual dimorphism has decreased during our evolution from an ape-like ancestor. Darwin proposed, and it is now accepted, that sexual dimorphism evolves by means of >sexual selection.

Sexual selection, a form of selection under which an individual has more offspring than average because of being able to acquire more mates. It was first described by Darwin, who though it both important and paradoxical. For him the puzzle was that by means of sexual selection a trait could develop that was not actually 'useful' to the species at all. The classic example is the peacock's tail, a spectacular and successful piece of sexual display but not a help to the bird in the business of getting its food or escaping predators. However, according to Darwin's correct view that what counts is the number of surviving offspring, a bird that gains more mates is more evolutionarily successful than another, regardless of their general competence in the environment. Darwin identified two components of sexual selection, and called them 'the power to charm the females' and 'the power to conquer other males in battle'. The latter leads to the evolution of offensive weapons in males, to be used not for hunting or defence but against other males of their own species (again a situation that does not on the face of it seem to be for the 'good' of the species).

Shuttle vector, a genetically engineered hybrid vector that can function in two different hosts. A plasmid that consists of DNA from both *Escherichia coli* and yeast is used for gene cloning with eukaryote genes. The gene is cloned in *E. coli*, then the plasmid is isolated and transferred to yeast which, being a eukaryote (albeit a primitive one), is more easily able to express eukaryote genes.

Sibling (sib), a brother or sister. Siblings have the same two parents; half-siblings have only one parent in common.

Sickle-cell anaemia, a disease in which the red blood cells are malformed and fragile and break up faster than they can be formed; this produces severe anaemia, which often proves fatal. Sickle-cell anaemia provides the classic example of Mendelian inheritance in humans, as the whole story is known – from the mutant base-change in the DNA to the evolutionary forces that keep the gene at a high frequency in many populations (*see* polymorphism).

Red blood cells contain a protein called haemoglobin, whose normal form is haemoglobin A (for simplification, the normal allele can be called Hb^A). There are several abnormal haemoglobins, the one involved in sickle-cell anaemia being haemoglobin S (allele Hb^S). Homozygous people (Hb^S/Hb^S) produce only haemoglobin S, which tends to form semi-crystalline lumps in the red blood cells, giving them a distorted shape (like a sickle, hence the name). Heterozygotes (Hb^A/Hb^S) produce both kinds of haemoglobin, and although they sometimes have mild anaemia they are generally healthy; they are said to have *sickle-cell trait*.

Analysis of the structure of the haemoglobins showed that the difference between the A and S forms is just one amino acid out of a chain that is 170 amino acids long, and this in turn is due to the >codon in the DNA changing from G-A-A to G-U-A. Thus the change of a single base in the DNA can make the difference between a person being healthy or having fatal anaemia.

It was discovered that people who were Hb^A/Hb^S heterozygotes were significantly less susceptible to malaria than people with only the normal haemoglobin A. It seems that the *Plasmodium* parasite finds the sickle-trait red blood cells more difficult to colonize and this provides the explanation for why a gene causing such an ill-effect as sickle-cell anaemia should occur at frequencies as high as 40 per cent in some tropical countries. Homozygous persons Hb^A/Hb^A are likely to die of malaria (if untreated), homozygotes Hb^S/Hb^S are likely to die of sickle-cell anaemia, and heterozygotes Hb^A/Hb^S have the best chance of survival. When heterozygotes have offspring, they keep both the Hb^S and the Hb^A gene in circulation. This situation, not rare in population genetics, is known as >heterozygous advantage. The Hb^S gene is only advantageous in areas with endemic malaria, and this can be seen in its distribution: Hb^S occurs in tropical Africa (with gene frequencies of about 15 per cent in West Africa, 40 per cent on Mount Ruwenzori in Zaïre, and up to 20 per cent in Madagascar, to take some examples), in Greece and Sicily, and in the southern extremes of the Arabian peninsula. It also occurs in black people of African origin, with an estimated 10 per cent frequency in the United States, where there was a notoriously ill-thought-out scheme in the 1970s for the >genetic screening of black people for sickle-cell anaemia.

Significance, the threshold level of >probability at which the experimenter decides to accept the results of statistical analysis. This will vary according to circumstances, but $P = 0.05$ (1 in 20) would be a minimum.

Silent mutation, any >mutation that does not have a detectable effect. *See* neutral allele.

Silk, the fibre made by some insects (for cocoons) and spiders (for webs). Silk consists mainly of a protein called fibroin which contains two amino acids: alanine and glycine. The silk worm, *Bombyx mori*, produces the greatest quantities of the fibre, and a curious genetic mechanism enables it to do so. The moth has only a single copy of the gene for fibroin, and it would not be physically possible for this to be transcribed fast enough to make silk at *Bombyx*'s speed; instead, in the silk-gland cells the entire genome is replicated about 18 times to give the extra copies of the gene. The more usual mechanism for coping with localized demand for high output from a gene is >gene amplification.

Skin colour. *See* pigmentation.

Smell, a sense of great importance to the many animals that communicate by means of >pheromones. In humans the sense of smell is much weaker than in other mammals. It is supposed that in human evolution sense has been sacrificed to the anatomical adjustments required for the enlarged brain, but this seems odd, given that humans still have physically quite as much space for the apparatus, external and neurological, as better-smelling animals have. Perhaps humans have a better smelling ability than they think they do, the sense of smell being under-used in human societies; but that would be odd too, given the advantage to every human group to make use of whatever resource it can. Maybe the sense of smell became disused after humans began to walk upright, when it became difficult to get the nose to the ground where most of the smells are. There is genetic variation among humans in the ability to smell particular things, and a curious example of this is that not everyone can smell every colour of that fragrant flower the freesia.

Smith, Hamilton (1931–), American molecular biologist; professor of molecular biology at Johns Hopkins University, Maryland since 1981. Smith made a crucial contribution to the technical repertoire of genetic engineering by isolating and identifying numerous restriction endonucleases in >*Escherichia coli*. These are the enzymes that can cut a >DNA molecule at one exact site. Smith and two of his colleagues shared the Nobel prize for physiology or medicine in 1978.

Snails, a group of molluscs consisting of thousands of aquatic and land species. The latter are hermaphrodites. The land snail *Limnaea* provides the best example of a >maternal effect.

Social Darwinism, the social and political ideas that were developed, in the second half of the nineteenth century, from Darwin's theory of evolution by natural selection, by Herbert >Spencer and others. Spencer believed that human societies evolved as species do, from 'undifferentiated hordes' to military societies and ultimately to industrial societies. It was also believed that the 'struggle for existence' described by Darwin as happening in nature was mirrored by the struggle for survival within human societies, so that social and economic inequalities were viewed as not only justifiable but necessary. *See also* pp. 40–1.

Social insects, a term covering the various orders of insects that live in organized colonies: ants, bees, wasps (some species), and termites. They are interesting not only because of their astonishingly elaborate 'societies' (including in the case of some ants the keeping of other species of ants as 'slaves') but also for their peculiar mechanism of sex determination.

The sex of an ant, bee or wasp does not depend on its sex chromosomes (they appear not to have any) but on whether the individual is haploid or diploid. The diploid insect is female; if she is fed suitably she becomes a queen and is fertile. She lays eggs; if these eggs have been fertilized they are diploid, and are therefore female, and they can either be fed to become queens or they can be left to remain sexually immature and infertile (these are the workers of the colony). Unfertilized eggs are haploid, and are male; every male therefore inherits all his genes from his mother. Males become fertile seasonally, producing gametes by mitosis rather than meiosis (their cells being haploid already). The advantage of this system is that the sex ratio in the colony is not fixed but can fluctuate to meet the environmental demands upon the colony, as the number of eggs fertilized can be raised or lowered.

Termites, though also fully social, have the ususal sex-chromosome mechanism for sex determination. There are kings as well as queens in termite colonies, and all termites are diploid.

Sociobiology, 'the systematic study of the biological basis of all social behaviour'. That is E. O. Wilson's definition in his book *Sociobiology: The New Synthesis* (1975). The essence of sociobiology is the belief that all behaviour patterns are genetically determined, and have evolved through natural selection; sociobiologists are thorough-going Darwinians. Among the questions that sociobiology has addressed are >territoriality, >aggression, mate selection, and parental behaviour. Its most controversial topic has been >altruism, for which there does not at first sight seem to be a plausible explanation in terms of natural selection.

Criticism of sociobiology has been strongly voiced, and is of two types. Firstly, geneticists have criticized it because of its reliance on >kin selection, the mathematics of which are only tenable in the case of very closely related kin. Secondly, anthropologists, psychologists, sociologists and others have criticized it because of its oversimplistic view that human behaviour can be thought of as consisting of a series of separate items determined genetically. The sociobiological explanations of homosexuality and incest have come under particularly heavy fire for this. Partly in response to the arguments of social scientists, some sociobiologists (and especially those in Britain) have stated that human social behaviour is a special case, with cultural inheritance as important as biological inheritance (*see* exogenetic inheritance). Richard Dawkins, in *The Selfish Gene* (1976), has proposed the concept of the >meme in order to maintain the idea of selection as the determinant of social behaviour but without being committed to genetic determination. A third type of criticism, that sociobiology is unacceptable because it is intrinsically fascist when applied to humans, is not a scientific argument but a political one. *See also* game theory.

Somatic cell, any cell in the body of an organism other than its >gametes, the sperm in its testes or the eggs in its ovary.

Somatic mutation, a mutation that occurs in a chromosome within a somatic cell. Such a mutation is not passed on to the individual's offspring, but it will be perpetuated in all cells descended from the one in which the mutation occurred. Some cancers are the result of a somatic mutation (*see* Philadelphia chromosome, and pp. 33–4).

Southern blotting, a technique for separating and identifying fragments of DNA. The DNA sample, perhaps the whole of a genome or a chromosome, is digested with restriction endonuclease and the fragments separated by electrophoresis in a gel. DNA is transferred (or blotted) on to a membrane filter, usually made of nitrocellulose or nylon. Specific sequences of DNA are then found by adding a radioactively labelled probe of RNA or DNA which forms a heteroduplex where the sequence is complementary. After excess probe has been washed away, the bound sequences of DNA can be detected by autoradiography (exposing a photographic plate to radiation from the sample) or scintillation counting (measuring the radiation with a device that shows particle collisions as flashes of light). Southern blotting is called after its inventor, but the parallel technique for RNA – >Northern blotting – is not.

Spacer DNA, a non-coding sequence of DNA between sequences that are gene-coding. An example is found in the gene for >ribosomal RNA, which has spacer sequences to separate the different RNA molecules that will eventually develop.

Spallanzani, Lazzaro (1729–99), Italian priest and physiologist. Spallanzani was the most brilliant experimental biologist of his day. He discovered that animals consume oxygen and excrete carbon dioxide (though these gases had not then been identified). He carried out experiments to disprove the idea of spontaneous generation; his sealed jars of previously boiled broth did not go mouldy, so he deduced that the agent causing the mould was something in the air (his experiments were very like those later carried out by Pasteur). By filtering semen, he discovered that spermatazoa are necessary for fertilization, and he carried out the first artificial insemination (on a spaniel bitch). Spallanzani investigated the circulation of the blood, and digestion (using himself as a guinea pig). He also researched the ability of bats to fly in semi-darkness without collisions, and realized that this was not a visual skill (the fact that bats use sonar was not discovered until 1941).

Speciation, the process of formation of a new species. Although it can refer to phyletic evolution, in which one species transforms itself into another over time, it more usually refers to cladogenic evolution, the splitting of a single species into two or more contemporary ones. 'Every species has come into existence coincident in time and space with a pre-existing closely allied species', wrote Wallace. Some mechanism of isolating the incipient species is needed (*see* isolating mechanism).

Species, a group of related organisms, all of which share the same gene pool and are potentially able to interbreed (producing fertile offspring). The species is the basic unit in classification and in the study of evolution. The test of whether two forms are able to interbreed cannot of course be applied to forms that are only known for their succession of each other in the fossil record; these have to be assigned to species rank using the same sort of criteria that are used for grouping species together into the same genus, etc. Every species has been given a two-part Latin name (*see* binomial nomenclature). The total number of species at present known to science is just under 1.4 million (about 1 million insects and other invertebrates, 250,000 vascular plants, 74,000 fungi and algae, 37,000 microorganisms, 19,000 fishes, 9000 birds, 9000 reptiles and amphibians, and 4000 mammals). It has been estimated that there may be as many as 30 million species, two-thirds of them tropical and the great majority of them insects.

Spencer, Herbert (1820–1903), British philosopher and sociologist. Spencer published a theory of evolution before the Darwin–Wallace one came out, but his was essentially the same as Lamarck's idea of evolution through the inheritance of acquired characteristics. Following the publication of *On the Origin of Species* in 1859, he became one of the staunchest champions of the theory of evolution by natural selection, though he did its cause a disservice when he coined the phrase 'the survival of the fittest', which has since been much misunderstood. Spencer expanded the evolutionary idea to include societies, postulating that in the same way as creatures evolve from simple to more complicated life forms so do societies evolve from the simple to the complex. He also drew analogies between the structure of an animal's body and the structure of a society. These ideas, and many more that favoured a rational rather than a religious view of humanity, were set out in a vast work, *Synthetic Philosophy*, published in ten volumes (1862–93). Spencer's ideas were an important starting point for sociology. *See also* Social Darwinism.

Sperm, the >gamete produced by male animals, insects, etc. (The equivalent in a plant is a pollen grain.) A spermatozoon (pl. spermatozoa) is the same thing. A human sperm has a head about 2 μm long, a mid-piece about three times longer, and a tail some 50 μm long. About 200 million sperm are present in each ejaculation. After they were first seen by van >Leeuwenhoek, it took a long time for the true role of sperm to be discovered. They were for a long time believed to contain a >homunculus (*see* pp. 4–5) and as recently as 1833 were thought to be a species of protozoa, with nothing to do with fertilization.

Sperm bank, a collection of sperm frozen and stored in liquid nitrogen, to be used for artificial insemination. Sperm banks for cattle have been routine for many years. The idea of setting up sperm banks for the genetical improvement of humans was strongly recommended by the pioneer American geneticist H. J. >Muller, whose work on >mutations had led him to worry that the quality of the human gene pool was declining because of the effect of medical science in preserving sub-optimal >genotypes; he was also worried about the mutational damage being caused by exposure to X-rays. The difficulty of deciding which men should be chosen to have their sperm in a bank is well illustrated by the fact that Muller was for a time keen on Communism and thought that Lenin was a great man and a prime candidate for the sperm bank; he later changed his mind completely. There are sperm banks which are used for artificial insemination in the treatment of infertility; the donors are popularly believed to be mostly medical students, and although very brief details on them are known to the doctor carrying out the treatment (notably their colour), their anonymity is strictly preserved and they cannot in law be held responsible for the paternity of tho child.

Spermatogenesis, the process by which a male makes gametes, i.e. >sperm. The starting point is in the tubules in the testis, which have a lining of cells called *spermatogonia*; these divide by >mitosis to form *spermatocytes*, which are also diploid cells. These go through the >meiosis divisions to produce four haploid *spermatids* each (all genetically different). The spermatids mature into sperm (properly, spermatozoa). The process of spermatogenesis goes on continuously in humans but is seasonal in most other species. Sperms that are not used in an ejaculation are retained in the testicle and liquefied.

Spina bifida, a congenital condition in which the vertebrae are imperfectly formed, leaving a gap through which the membranes of the spinal cord stick out. In many cases, surgery can repair some of the damage, though a child may need many successive operations. About 3 per 1000 babies in the British Isles are born with spina bifida, which is more common in northern and western England and in Eire. There seems to be some genetic predisposition to the condition, probably involving polygenic inheritance, but the mother's vitamin intake may also be a contributing factor.

Spindle, a structure made of minute fibres that appears in the cell's nucleus and is involved in organizing the movement of chromosomes during cell division (*see* meiosis; mitosis).

Splicing, the technique for joining two molecules of DNA together, for making >recombinant DNA. The basic procedure is to cut both the DNA molecules with a restriction endonuclease that makes a staggered cut in the sequence, leaving a complementary-reading tail (e.g. one that reads AATT; this is complementary to itself, in that A must always pair with T). The two sorts of DNA are then mixed together, and since both have the same complementary tails they stick together, the joins being made by the enzyme DNA ligase. This is the technique used in gene cloning, in which the gene required to be cloned is spliced to the DNA of a plasmid, which can then be taken up by a bacterial cell.

Spontaneous generation, the belief that living matter can arise from non-living matter. This was a universal belief up till its final disproof by Louis >Pasteur in the 1860s. *See also* p. 7.

Spontaneous mutation, a mutation that arises in natural conditions, as opposed to one that has been induced with a mutagen.

Spore, a unicellular reproductive unit that becomes detached from the parent cell and develops into a new individual, either after fertilization by another spore or by >vegetative propagation. In plants, the spore is the haploid product of meiosis (*see* alternation of generations) and is comparable to a gamete in an animal.

Sporophyte, the diploid phase of a plant's life cycle. In higher plants it consists of the plant itself as visible to the naked eye. *See* alternation of generations.

Sport, an individual with a visibly different phenotype to the normal, due to a >mutation. The word is from the pre-Mendelian era, having been used in this sense from about 1840, and is now only used in horticulture. A bud sport in botany is a shoot that is different from the plant it grows on, following a >somatic mutation.

Stabilizing selection, the type of natural selection that acts to keep the phenotype of a population stable around a mean. Extremes are selected against, and the amount of underlying genetic variance is reduced. Stabilizing selection only occurs in a population in a constant environment. *See also* directional selection; disruptive selection.

Standard deviation, a measure of how widely the individual observations in a given sample vary from the mean. It is calculated as the square root of the variance. As a rule of thumb, in a normal distribution 95 per cent of the observations lie within two standard deviations either side of the mean.

Start codon, the codon in messenger RNA that signals the point at which >protein synthesis should start. It is AUG, which codes for the amino acid methionine. *See* genetic code.

Stasis, the existence of a species in an unchanged form over a long period of time. An unchanging environment is necessary. The coelacanth, a lunged fish ancestral to the amphibians, is an example, being little changed since its first appearance some 350 million years ago.

Statistics, a set of techniques for analysing numerical data. Statistical analysis cannot say whether a hypothesis actually is true or false, merely that it is more or less likely to be true or false. The science of statistics in general, and especially its use in biological problems, was pioneered by >Fisher in the 1920s and 1930s.

Sterility, inability to produce viable gametes. This may be due to absence or defectiveness of gonads, or there may be a chromosomal difficulty that prevents meiosis taking place (e.g. in the case of triploids or interspecies hybrids such as mules). *See also* infertility.

Sticky end, the end of a DNA molecule where one strand extends beyond the other by a few bases, as opposed to one which ends flush (*see* blunt end). Two sticky ends can join together if their sequences of bases are complementary.

Stop codon. *See* nonsense codon.

Structural gene, any gene that codes for a gene product, in contrast to a gene that is a spacing or regulatory sequence (*see* spacer DNA; regulator gene).

Student's t-test. *See* t-test.

Sturtevant, Alfred Henry (1891–1970), American geneticist. Sturtevant worked in T. H. >Morgan's 'fly room' at Columbia University, studying the genetics of *Drosophila*. In 1912, while still an undergraduate, he had the idea for >genetic mapping, based on the frequency of crossing-over between genes on the same chromosome. He went on to discover that the effect that genes at one locus have on genes at another locus depends on the distance between the two loci (i.e. position effect). Sturtevant also found that where one chromosome contained an inverted sequence (i.e. an inversion) no crossing-over took place. He did research on various other animals apart from *Drosophila*: his papers on coat colour in horses (based in part on data from his boyhood on his father's farm) and on left- or right-coiling snail shells (*see* maternal effect) are classics.

Substitutional load, the part of the >genetic load of a population which consists of 'old' alleles in the process of being replaced with new better ones.

Substrate, the molecule or compound upon which an >enzyme acts. Each substrate is specific to its own enzyme, and vice versa.

Succession, principle of, the principle that, in the geological record, simpler life forms are replaced in time by more elaborate ones, and that the sequence in which this occurs is the same from one location to another. This was established by the British geologist William Smith in 1799 (*see* p. 14), and was an important part of the evidence that was accumulating in favour of the evolution of species. A related finding was that the gaps between present-day life forms can often be filled by intermediate forms that existed in previous ages. *See also* distribution, principle of.

Supergene, a group of genes in very close linkage, with crossing-over between them suppressed so that they are inherited as a single unit. Supergenes have been found in butterflies, where the several genes involved in mimicry have become closely linked.

Suppressor mutation, a mutation that acts to restore a function that had been lost after an earlier mutation. It may act within the same gene (possibly by changing back a frameshift) or in another gene, by coding for a tRNA that will read the mutated nonsense or missense codon as the normal codon.

Surrogacy, the bearing of a child by one woman to be brought up by another. Let the woman who carries the baby be Mrs A (she is usually married) and the couple wanting to the baby Mr and Mrs B. In *partial surrogacy*, Mrs A is made pregnant by Mr B, either by sexual intercourse or artificial insemination. The baby therefore has half its genes from Mr B and half from Mrs A, and is as closely related to Mrs A as any other baby she might have. In *full surrogacy*, *in vitro* fertilization is used: an egg from Mrs B is fertilized by a sperm from Mr B, and the embryo is implanted into Mrs A, who has no genetic relationship to it at all (*see* embryo transfer). The baby is entirely genetically the offspring of Mr and Mrs B who will bring it up. Partial surrogacy has a long history, the earliest recorded instance being in Genesis 16, and as it can be informally arranged without involving the medical profession it may well be more common than official statistics suggest. Full surrogacy became an issue at the beginning of the 1980s, when *in vitro* fertilization technology became reliable. The British Medical Association's view in 1983 was to doubt whether full surrogacy would ever be acceptable. A succession of sensationalized cases soon followed, including one in South Africa in which a woman gave birth to her own grandchildren – triplets born from her daughter's eggs. With the passing of the Human Fertilization and Embryo Research Act both partial and full surrogacy are now legal in Britain, though with various safeguards including a prohibition on payment to the surrogate mother other than to cover her expenses.

Switch gene. *See* oligogene.

Symbiosis, the arrangement in which two species live closely together. It may be for their mutual benefit (*mutualism*), as with the species of ants that keep aphids as 'cows'; the benefit to the 'cows' is that they are protected against predators. Such a system requires co-ordinated evolution of the two species. In *commensalism*, two species share a common resource without affecting one another either adversely or favourably; this might be two species that live in the same burrow. *Inquilinism* is a symbiosis in which are species lives inside the body of the other but without causing the host any harm; this is the relationship between humans and *Escherichia coli*. The most usual form of symbiosis is *parasitism*.

Sympatry, the existence in the same area of two or more related species.

Synapsis, the pairing of the homologous chromosomes during the first prophase of >meiosis.

Syndrome, a medical term denoting a number of symptoms that occur together (the word comes from the Greek *syn*, 'together', *dromos*, 'a race', 'a running [together]'. The cause may be infective, environmental or genetic. A syndrome caused by a single gene is an example of >pleiotropy, but many genetically caused syndromes result from chromosome abnormalities, not single defective genes.

Systematics, the study of the relationships of the different groups of living things. The term is more or less interchangeable with >taxonomy, but is more old-fashioned. *See* classification.

T

T lymphocytes a group of white blood cells involved in the immune system. They have on their surfaces receptors that recognize histocompatibilty antigens on the surfaces of cells, so that they bind on to these antigens on any foreign cell that enters the body. They are therefore the main part of the body's response to transplants (*see* rejection). T lymphocytes (so called because they originate in the thymus) recognize the antigens by sequence (unlike >B lymphocytes).

T4, one of the most intensively studied of the bacteriophages. It only attacks >*Escherichia coli*. The genome is double-stranded DNA, linear, but with matching ends that can join up to make a circular DNA molecule after entry into the host cell.

Tailless, a Mendelian allele occurring in mice. The inheritance of the tailless gene was, at first, a great puzzle to geneticists, as the trait appeared in succeeding generations as a Mendelian allele would, but with unusual pedigrees. Litters would be either all normal, or all tailless, or one normal and the rest tailless. Never the expected Mendelian ratios of 1:1 or 3:1. And was taillessness dominant or recessive? – two normal parents might have tailless offspring (which rules out dominant), but two tailless parents might have normal offspring (which rules out recessive). The penny finally dropped in the mind, not of one the researchers, but of a laboratory technician: the gene is not t = tail*less*; it is t = tail-*eater*. What really happens is that the litter can be *(a)* all normal (no *tt* pups so no tails eaten), (*b*) all tailless except one (one *tt* pup that eats all its siblings' tails but not its own) or *(c)* all tailless (two or more *tt* pups that eat each other's tails as well as those of their siblings). The moral of the story is that what the scientist believes to be the phenotype may be nothing of the kind.

Tandem duplication, a form of duplication in which part of a chromosome duplicates itself, the second copy being inserted in line with the first.

Taste blindness, a condition in which an individual is unable to taste the chemical compound phenylthiocarbamide (PTC). It has a slightly bitter taste, and is found in some members of the cabbage family. The inability to taste PTC is inherited as an autosomal recessive. Non-tasters comprise 25–30 per cent of the population in northern Europe but only 7 per cent of the Japanese and 1 per cent of the South American Indians. There must be some selective advantage to keep the gene in populations at such high frequencies, and it has been suggested that, as PTC is chemically quite close to some substances that affect thyroid function, it may be something to do with the prevention of goitre (enlargment of the thyroid). It is believed that chimpanzees have a similar gene. Taste blindness was discovered by chance in 1931.

TATA box, a sequence found in eukaryote genes. Part of the >promoter region, it occurs about 35 base-pairs upstream from the point at which RNA transcription is to start. Prokaryotes have a >Pribnow box.

Tatum, Edward Lawrie (1909–75), American microbiologist. Tatum provided the microbiology know-how in the joint experiments that he and George >Beadle carried out on the bread mould *Neurospora crassa*. In 1941 they showed that one gene codes for one enzyme. Tatum next worked with Joshua >Lederberg on genetic recombination in *Escherichia coli*, and their work was the foundation of the techniques of genetic engineering. Tatum, Beadle and Lederberg shared the 1958 Nobel prize for physiology or medicine.

Taxon (pl. taxa), any group of living things, from the viewpoint of >classification. The key taxon is the species; the higher taxa are:

- *For animals*: genus, family, order, class, phylum, kingdom.
- *For plants*: genus, tribe, family, order, class, division, kingdom.

Lower taxonomic divisions into subspecies and varieties can also be made.

Taxonomy, the arrangement of taxa into a system of >classification. The outline of the scheme of classification was laid down in the eighteenth century by >Linnaeus, Cuvier and others (*see* pp. 13–14), using only anatomical features (and before evolution was accepted, so there was no element of keeping ancestrally related forms together). A modern technique is numerical taxonomy, in which all attributes of the phenotype – metric, biochemical or whatever – are given equal weighting, and groupings are worked out by computer.

Tay–Sachs disease, a human disease in which the baby is born normal but suffers mental retardation after the age of three months, followed by blindness and paralysis; death usually occurs at the age of about two years. It is inherited as an autosomal recessive, the biochemical defect being an inability to produce the enzyme hexosaminidase A. Tay–Sachs disease occurs at a frequency of about 1 in 2500 births among people of Ashkenazi Jewish origin, but only about 1 in 100,000 in other populations (including Sephardic Jews). Carriers of the gene can be detected by a biochemical test. There was a highly successful programme of >genetic screening among the Jewish populations of Washington D.C. and Baltimore, Maryland in the United States to identify couples both of whom were carriers so that they could take appropriate action to avoid having an affected baby; this programme became the model for a number of similar schemes among Jewish people.

Telegony, the belief that the first male that mates with a female has a lasting effect on her subsequent offspring, even when these are fathered by a different male. It is difficult to understand how this was supposed to work: even with blending theories of inheritance (*see* pp. 7–9) one would not expect something in the male's contribution to 'stain' the uterus (or would one?). There was alleged evidence of telegony in the case of an Arab mare mated with a zebra, who later (and when mated to another Arab) produced foals that had bristly manes and some stripes on their legs; even Darwin accepted this proof. The experiment was repeated by a professor of veterinary science who bred many such crosses and found no evidence for telegony, but did confirm that quite a few pure-bred horse foals have stiff manes and striped legs anyway.

Many in animal-breeding circles still believe in telegony, though not officially; for instance the Kennel Club in Britain is perfectly happy to register pure-bred puppies no matter how chequered the reproductive history of the bitch. And even in the time when the zebra story was believed, it was not held to apply to humans: you could marry a rich young widow and take advantage of the money without worrying that her previous mate might reappear in your children. (The converse belief, that the first female he mates with affects all the subsequent offspring of a male, has never been so prevalent – and definitely did not recommend itself in the case of humans.)

Teleology, the belief that there is evidence of purpose in the universe. In the context of biology, it is particularly taken to mean the idea that evolution (especially when seen as the path from the primitive origin of life to the glorious culmination in *Homo sapiens*) has not been purely mechanistic but has been purposively achieved. This is of course not the point of view of most modern biologists, who are quite satisfied with the mechanistic process of evolution by natural selection working on a trial-and-error basis with the raw material provided by random mutations.

The teleologists' 'purpose' has to come from one of two sources. One could be the will of the organisms themselves as they evolve. H. G. Wells and Julian Huxley wrote scathingly of 'the difficulty of ascribing even rudimentary purpose or foreknowledge to a tapeworm or a potato, or collective aspiration to the tapeworm race or the potatoes of the world'. The other possible source of purpose is a Divine Creator, the existence or non-existence of which is not provable. Biology cannot refute the proposition that natural selection is the tool of a Creator (*see* creationism). Another point is that a modern biologist would not see evolution as a process ending with the arrival of humans; if God made us in His own image, then that image must include every step of the evolutionary journey from the primeval sludge to modern humanity and beyond.

Telocentric chromosome, one with the >centromere at the extreme end.

Telophase, one of the phases of the two types of cell division, >meiosis and >mitosis. *See* diagrams at meiosis and mitosis.

Teratogen, any environmental factor that causes an embryo or foetus to become deformed. The damage is not genetic but congenital (*see* congenital defects). Most pharmacological drugs are small molecules that can pass through the placenta into the unborn baby's bloodstream, and are thus always suspected of being teratogenic until proved otherwise. *See* thalidomide.

Terminal transferase, an enzyme that synthesizes a 'tail' on to a DNA molecule. It is used in genetic engineering to add a poly-A tail (one that consists only of repeated As) to the end of one DNA molecule and a poly-T tail (one that consists only of repeated Ts) to the end of another so that they join up.

Territoriality, the behaviour by which a group or individual claims rights over a certain space. Almost all vertebrates, and many of the more advanced invertebrates, recognize specific areas as their own. The widest area is the *total range*, the maximum area (or volume, for all these concepts apply to underwater and aerial habitats as well as purely terrestrial ones) that the individual visits in its lifetime. The *home range* is the area known thoroughly by the individual and visited regularly by it. This may or may not be the same as the *territory*, defined as the area or volume that an individual, pair or group actively excludes others from, by marking (*see* pheromones), display or aggressive behaviour. The key resource that is guarded in the territory may be food supply, shelter, nesting sites or space for sexual display. The territory can change with time, for example during seasonal migration. Sociobiologists hold that territoriality, as other behaviour, is genetically controlled.

Testcross, a cross between an individual with the dominant phenotype but of unknown genotype and an individual homozygous for the corresponding recessive, the cross being made in order to find out the genotype of the first individual. The cross is either $AA \times aa$ or $Aa \times aa$, and the phenotypes of the offspring will show which it is, by being either all dominant-type or 50 per cent dominant and 50 per cent recessive. A testcross can involve more than one locus. *See also* backcross.

Testicular feminization syndrome, a condition in which the person has the chromosome complement of a normal male (i.e. 46, XY) but owing to a lack of receptors for the male sex hormone testosterone, does not develop normal testicles. The external genitalia look female, and a boy with this syndrome will be brought up as a girl. The syndrome is inherited as an X-linked recessive, and is the only example of a gene on the X chromosome that affects sexual development (*see* sex determination).

Tetrad, the four homologous chromatids of a chromosome pair as they lie together during >meiosis. The term is also used for the group of four haploid cells which are the product of meiosis. *See* tetrad analysis.

Tetrad analysis, a technique for studying linkage and recombination, using the tetrad of meiotic products in a fungus such as *Neurospora crassa*. The tetrad of four haploid cells lies in a line in the ascus, and a mitosis division takes place so that there are eight cells (>ascospores), lying in four pairs, each pair representing one product of meiosis. Each of the ascospores can be grown separately (as *Neurospora* is in the haploid state for most of its life cycle) and the presence or absence of particular alleles noted.

Tetraparental mouse, a mouse produced by mixing *in vitro* two different blastulas; also known as an 'allophenic mouse'. The resulting mixed ball of cells is usually capable of developing into an embryo, and can be reimplanted into a mouse for the remainder of its prenatal development. The mouse has four parents instead of two, and these four can all be of different strains. However, each of its cells has only two parents, being derived from one or other of the original 'ingredients'. The mouse is therefore an artificially produced >chimaera.

Tetraploidy, the condition of having four copies of each chromosome. *See* polyploidy.

Thalassaemia (from the Greek *thalassa*, 'the sea'), a disease in which the red blood cells are malformed and fragile and, as a result, break down faster than they can be produced; this results in severe anaemia, which often proves fatal. Thalassaemia is probably the commonest single-gene disorder in the world, with high gene frequencies right across the Old World, from Africa to New Guinea, between latitudes 40°N and 15°S.

The biochemical problem is a defect in either the alpha or beta chain of the haemoglobin molecule. In alpha-thalassaemia, the homozygous condition is fatal at birth (*hydrops foetalis*). In beta-thalassaemia, homozygotes suffer *thalassaemia major*, a severe anaemia requiring regular blood transfusions (untreated patients die at about two years old), while heterozygotes have *thalassaemia minor*, a less severe anaemia but one that is still fairly debilitating. There are some 30 different mutations (frameshifts, deletions, insertions or inversions) that can give rise to beta-thalassaemia. As with >sickle-cell anaemia, the evolutionary justification for populations to have a high frequency of a gene for this disease is that heterozygotes are protected to some extent against malaria, which until recently was one of the major killers in the areas where thalassaemia occurs.

A simple blood test (carried out routinely in many countries) can detect whether a pregnant woman is a carrier of the thalassaemia gene, and couples who are both carriers (the father can also be tested in this way) can be advised that they are at risk of having a thalassaemia major baby. Prenatal diagnosis of the babies of such couples, however, is not so simple, because the babies do not start to make haemoglobin beta-chains until a late stage of gestation; but in some families, it may be possible to use genetic markers to tell with a high degree of probability whether or not the baby has inherited the thalassaemia gene from both parents. In places such as Cyprus and Sardinia where the disease is very common, schemes for the prevention of births of thalassaemia major babies have given interesting examples of the cost-effectiveness of this policy (*see* pp. 42–3).

Thalidomide, a sedative drug that was put on the market in 1957 after extensive testing on four different species of mammals, including testing for birth defects. It was sold in Europe without prescription, but was withdrawn in 1961 when its association with phocomelia (a deformity in which the hands and/or feet are joined to the body by very short limbs or none at all) was noticed. Some 8000 babies worldwide were born with (and survived) this defect (congenital, not genetic) after their mothers had taken thalidomide during the first three months of pregnancy; many others with extensive abnormalities died before birth. The reason for mentioning thalidomide in a book on genetics is that the story emphasizes the genetic differences between species: rodents can take thalidomide with impunity, we cannot.

Theophrastus (*c.* 371–288 BC), Greek philosopher. Theophrastus studied under >Aristotle, and took over the school at the Lyceum from him when he retired in 335 BC. He lectured on as wide a range of subjects as Aristotle had done, but he is chiefly remembered for his book *Historia plantarum* (*On the History of Plants*), in which he describes the roles of flower, fruit and seed, and also states that in some plants there are female flowers that must be fertilized by a male (an idea a long way ahead of its time, which was not fully described until Camerarius's work in 1694).

3′ ('three prime'), one end of a molecule of RNA or DNA, the name referring to the number of the carbon atom in the sugar that comes at the end (joined to a phosphate). The other end is known as >5′, the links between the sugars throughout the molecule being between the 5′ and the 3′ carbons. 3′ is the 'tail' end. (*See* the diagram at >replication.)

Threshold trait, a trait in which the underlying cause has >continuous variation but in which the phenotype is expressed only at a certain level. An example is cleft palate, in which the disorder is due to a delay in several interacting developmental processes. Up to a certain point this delay is not harmful, but once that critical point is reached one process – fusion of the palate – misses its proper place in the schedule and cannot occur. In threshold traits, the determining factors are both genetic and environmental, so patterns of inheritance are very hard to work out.

Thymine (abbreviation: T), one of the four nitrogenous bases which form the core of DNA (but not RNA, where its place is taken by >uracil). It pairs with adenine (A).

Tissue culture, the maintenance *in vitro* of a line of cells taken from a tissue of an organism (*see* cell line). Depending on the type of cell, a culture may be able to go on indefinitely, with cells dividing from time to time. Cancer cells (or normal cells that have undergone transformation in culture) will go on dividing infinitely often, but most differentiated cells seem to lose the ability to divide after 30–40 divisions; this is known as the >Hayflick phenomenon, and it is not known how it is related to the normal process of ageing.

Tissue typing, the identification of >histocompatibility antigens in donor and recipient prior to transplant surgery.

Tortoiseshell cat, a female cat whose coat colour is made up of random patches of orange and black. The genes for these colours are on the X chromosome, and when random inactivation of one or other X chromosome in each cell takes place (*see* Lyon hypothesis), either the chromosome with the 'black' gene or the one with the 'orange' gene is left active.

Trans configuration (also known as repulsion), one of the two possible arrangements for the alleles in an individual that is heterozygous for mutations at two linked loci. *Trans* (from the Latin, 'across') means that one of the mutant alleles is on one chromosome and the other is on the homologous chromosome, i.e. the two chromosomes are $\underline{a\ +}$ and $\underline{+\ b}$. The opposite is >*cis*.

Transcription, the process by which one strand of DNA is copied into a single strand of RNA; the first step in >protein synthesis. The RNA molecule is synthesized by RNA polymerase, copying the DNA message by following the base-pairing rules (*see* genetic code). The polymerase first recognizes the promoter region, and actual transcription begins at the start codon a few bases further downstream (going from the 5′ towards the 3′ end of the DNA strand).

Transduction, the transfer of genetic information from one bacterium to another when it is carried by a >bacteriophage.

Transfer RNA (abbreviation: tRNA), the type of small RNA molecule that assembles the amino acids for >protein synthesis. Each transfer RNA molecule is shaped roughly like a clover leaf, and consists of an anticodon at one end (which is complementary to the appropriate codon in messenger

T

RNA) and its specific amino acid at the other; there is also a ribosome binding site. In principle, there could be 61 different transfer RNA molecules, one for each 'sense' codon in the >genetic code, sharing the 20 amino acids among them; in practice, each cell contains 40–60 different ones.

Transformation. This has two meanings in genetics. *(1)* A permanent change in the genetic characteristics of one bacterium by exposure to DNA of a different origin. It was the phenomenon of transformation that led >Avery to the discovery that DNA was the molecule responsible for carrying genetic information. *(2)* A change in an animal cell in tissue culture so that it grows and divides in the same way as a cancer cell, possibly due to activation of a viral gene. The relationship of this phenomenon to cancer is not understood.

Transgenic organism, any organism that has been transformed (to however slight a degree) by having DNA from another species integrated into its genome. Note that a transgenic mouse – say with a single chimpanzee gene in its genome – is not a mouse/chimpanzee cross; it is still just a mouse. *See also* genetic engineering.

Transient polymorphism. *See* polymorphism.

Transition, a mutation in which a purine base (adenine, guanine) is replaced with another purine, or a pyrimidine (cytosine, thymine) with another pyrimidine. *See also* transversion.

Translation, the process by which the protein coded for in messenger RNA is built up, by transfer RNA and ribosomes. *See* central dogma; protein synthesis.

Translocation, a piece of one chromosome that has broken off and become attached to a different chromosome. There is usually no adverse effect if this happens in a somatic cell, as descendant cells have all the right genes even if some are in the wrong place (though sometimes the translocation does have a harmful effect; *see* Philadelphia chromosome). However, if the translocation happens while a gamete is being made, some of the descendant cells will have surplus genetic information and some will have what is in effect a deletion. *See also* Robertsonian translocation.

Transplants, organs surgically moved from donors to recipients. For the problems of the immunology, *see* HLA system; immunosuppression. *See also* graft.

Transposition, the movement of a >transposon or other movable sequence of DNA from one place in the genome to another.

Transposon, a movable genetic element similar to a >jumping gene. They are found in the genome of bacteria, and are involved in the resistance to antibiotics. Unlike jumping genes, transposons leave a copy of themselves in the original position when they move to a new site. They make it possible for antibiotic resistance to be spread very rapidly, not only within a single strain of bacteria, but from one genus to another. (*See* conjugation.)

Transversion, a mutation in which a purine base (adenine, guanine) is substituted for a pyrimidine one (cytosine, thymine), or vice versa. *See also* transition.

Trihybrid, a cross between individuals that are >heterozygous at three different loci. The genotype of these individuals being *Aa Bb Cc*, there are eight different gametes that each can produce – *A B C*, *A B c*, *A b c*, *A b C*, *a b c*, *a b C*, *a B C*, *a B c* – which leads to 64 (8^2) possible genotypic combinations in the offspring. The ratio of the phenotypes among the offspring is 27:9:9:9:3:3:3:1 – 27 showing dominant type for all three traits, 9 for each possible pair of dominant phenotypes, 3 for each single occurrence of the dominant, and 1 for the triple recessive. *See also* hybrid.

Triplet, three bases in sequence in DNA or RNA. It forms a 'word' in the >genetic code. *See* codon.

Triplets, three babies born together. The frequency with which triplets are born is the square of the frequency of twin births (with quadruplets the cube), though there is no theoretical explanation for this observation (*see* twinning rate). This rate is no longer the same, given the high frequency of >multiple births following the use of fertility drugs.

Triploidy, the condition of having three copies of every chromosome (a form of >polyploidy). Triploidy does not occur in nature as a permanent feature of a species, as triploid organisms have very low fertility. This is because at meiosis there are three chromosomes trying to find partners for a process that only works with two, so the division is usually a failure. However, triploid organisms are just as capable of normal growth as any others (mitosis is not impeded at all, since the process consists of each chromosome dividing itself in half, irrespective of any partner).

Artificially produced triploids can be commercially useful. One example of this is the banana, the trees of which are bred as triploids from stocks of a tetraploid and a diploid strain. The triploid hybrids grow well, but their seeds never develop properly, so that the inside of the banana is not full of inedible seeds as it is in a normal banana.

Trisomy, the condition of having three copies of one particular chromosome. This almost always produces some abnormality, as genes are deleterious when present in greater than normal numbers. In humans the best-known example is trisomy 21 (>Down's syndrome), but there are others that occur quite frequently, including trisomy 18 (Edwards syndrome; 1 birth in every 8000 live births, but ten times this number are spontaneously aborted) and trisomy 13 (Patau syndrome; 1 in every 20,000 live births, 20 times more often spontaneously aborted). Trisomy happens as a result of translocation or of non-disjunction of homologous chromosomes during meiosis.

Triticum aestivum, common bread wheat. Primitive (emmer) wheat was first domesticated in about 7000 BC. The present form is hexaploid, i.e. it has six of each chromosome, representing three sets each of different origin ($6n = 66$). One set is homologous with the wild wheat *T. monococcum* ($2n = 22$), and another with a related wheat-like grass *Aegilops squarrosa* ($2n = 22$), but the third has not been identified and may have come from a species that is now extinct. There are at least 20,000 varieties of *T. aestivum* in cultivation. The wheat used for pasta is a different species, *T. durum*.

tRNA. *See* transfer RNA.

True breeding, the phenomenon in which a strain of animals or plants produces offspring with the same genotype as its own. The term is most often used in connection with a recessive >allele – for example, Suffolk Punch horses which breed true for chestnut colour – but it could also be used for a strain homozygous for a dominant allele, or for genes at more than one locus.

Tschermak von Seysenegg, Erich (1871–1962), Austrian botanist; one of the three rediscoverers of >Mendel's laws. A member of a distinguished family of scientists, Tschermak specialized in agricultural botany, and his experiments aimed at producing higher-yielding varieties of corn led him to the independent discovery of the principles of heredity first described by Mendel in 1865. Tschermak found Mendel's long-neglected paper in 1900, at the same time as de >Vries and >Correns, who had both discovered the same principles through their own experiments. Tschermak was a very successful breeder of plant varieties: his *Aegilotriticum* wheat–rye cross was the first artificially produced hybrid between two different genera, and his rye variety 'Marchfield' is still the highest yielding in the world.

t-test, a statistical technique for finding whether the mean value of one sample is different from that of another sample. It is often used in genetics and ecology; an example would be a comparison of the weights of mice caught in one place with mice caught in another place. It is also known as Student's t-test, from the pseudonym of the mathematician W. R. Gosset who devised it.

Turner's syndrome, the condition in which a female has genotype XO, i.e. only one sex chromosome. A Turner's syndrome girl looks fairly normal, but with short stature and a wide ('webbed') neck; the ovaries are present but undeveloped so the she is not fertile. Although there is seldom any general mental retardation, there is often a defect in spatial perception. The single X chromosome can be inherited either from the mother or the father, but usually it is the maternal X, a normal ovum having been fertilized by a sperm that lacked a sex chromosome. If a Turner's syndrome patient inherits an X-linked recessive gene – e.g. for red–green colour blindness – she will have the abnormality, as there is no corresponding normal X chromosome to 'cover' the recessive (*see* X linkage). Turner's syndrome occurs in about 1 in 1500 births in Britain.

Twin studies, research using pairs of twins to study the effects of genotype and environment on human characteristics. Both monozygotic (identical) and dizygotic (non-identical) twins are used. The theoretical basis is that since monozygotic twins have identical genotypes, any differences between them are produced by the environment. Normally the environment is very similar for both twins, particularly as parents of twins often treat their children exactly alike, even to point of dressing them in identical clothes.

Coefficients of correlation for three traits, as determined by twin studies

	Monozygotic together		
	Newman et al.	*Shields*	*Burt*
Height	+0.932	+0.98 males +0.94 females	+0.962
Weight	+0.917	+0.79 males +0.81 females	+0.929
IQ	+0.881	+0.76 females	+0.925
	Monozygotic apart		
	Newman et al.	*Shields*	*Burt*
Height	+0.969	+0.82 males +0.82 females	+0.943
Weight	+0.886	+0.87 males +0.37 females	+0.884
IQ	+0.670	+0.77 females	+0.874
	Dizygotic together		
	Newman et al.	*Shields*	*Burt*
Height	+0.645	+0.44 females	+0.472
Weight	+0.631	*n.a.*	+0.586
IQ	+0.631	+0.51	+0.453

Dizygotic twins have (on average) half their genes in common, as do any other pair of siblings. If one assumes that the two kinds of twins experience environments that are to a similar extent identical (probably an oversimplification, though parents are usually careful to be even-handed with twins), the difference in the amount of similarity or divergence between twin pairs of both sorts can be taken as an indication of the effect of 100 per cent of genes in common versus 50 per cent common. This can be reinforced by looking at siblings reared together, who have the same amount of genes in common as dizygotic twins but a less-than-identical environment (due to differences in age, etc.).

The ideal situation (from the geneticist's point of view, never mind anyone else's) would be monozygotic twins separated at birth and raised entirely separately, preferably (again from the geneticist's point of view) in very different environments. The divergences between such pairs of twins measure the amount of influence that the environment is capable of exerting. Not many such pairs of twins exist: the classic separated-twin studies are those of Newman, Freeman and Holzinger (US, 1937) with 19 pairs and Shields (Britain, 1962) with 44 pairs; >Burt (Britain, 1966) claimed 53 pairs, but was later shown to have fabricated data on many of them (although some were genuine). The table above shows data from these researches; IQ is included (though it is now a highly suspect commodity) because it was one of the traits most keenly studied by this methods, and Burt's spurious data are included for interest.

Twinning rate, the frequency of twin births, as a percentage of all births. There is considerable variation between populations, with the lowest (0.65 per cent) found in Japan and the highest (4.5 per cent) in Nigeria, the average being between 1 and 1.5 per cent. The proportion of these that are monozygotic (identical) can be calculated from the fact that half the pairs of dizygotic twins (non-identical) are of unlike sex, and that there is a corresponding number of like-sex non-identical pairs; the excess of like-sex twin pairs can be assumed to represent the monozygotic twins. For example, if 50 pairs of unlike-sex twins are born per year, and 60 pairs of like-sex, one can deduce that the 'extra' 10 pairs are monozygotic. The birth rates of monozygotic twins is not very different from one population to another, the variation in twinning being due almost entirely to the rate of dizygotic twins.

There is some genetic basis for a tendency of a mother to have twins, but there is also a very strong maternal-age effect, with older women more frequently having twins (again, the difference being in dizygotic twins), and with birth order, later babies being more likely to be (once again, dizygotic) twins. *See also* triplets; quadruplets.

Twins. *See* dizygotic twins; monozygotic twins; twinning rate.

U

Ultraviolet light, electromagnetic radiation of a wavelength of about 400 to 4 nm, between the visible spectrum and X-rays. It is present in sunlight, and causes damage to the chromosomes in any cells that it reaches (*see* DNA repair). It is to protect the skin cells from ultraviolet rays that the brown pigmentation of human skin evolved (*see* melanin). However, humans need ultraviolet radiation in order to make vitamin D.

Uniformitarianism, the theory that the processes by which the physical structures of the Earth were formed are continuous and are still going on. The idea was first proposed in 1788 by the British geologist James Hutton (*see* p. 12), who was bitterly attacked because it was counter to religious teaching. Uniformitarianism was later championed by Sir Charles >Lyell (*see also* p. 15) and was part of the accumulating evidence in favour of a process of biological evolution.

Unisexual, having only one kind of sex organ; usually used of a plant that has flowers of only one sex (*see* self-sterility).

Upstream, further back on a DNA molecule, in respect of the direction in which the sequence is being read. *See* replication.

Uracil (abbreviation: U), one of the four nitrogenous bases which form the core of RNA (but not DNA, where its place is taken by thymine). It pairs with adenine (A).

V

V region, or variable region, part of the heavy and light chains of an >immunoglobulin molecule. It is this region that interacts with the antigen.

Vaccination, the artificial production of active immunity by the injection of a vaccine. The word 'vaccination' is derived from the Latin for 'cow', and was coined by Edward Jenner at the end of the eighteenth century to describe his idea of injecting patients with cowpox virus to protect them against smallpox (a very similar virus). A vaccine now means any suspension of dead or non-virulent viruses or bacteria used in immunization. *See also* Pasteur, Louis.

Variable number tandem repeats (VNTRs), the repeated sequences of DNA that vary from one individual to another and are the basis for >genetic fingerprinting.

Variable region. *See* V region.

Variance, a measure of the spread of values of observations around the >mean of a distribution. It is calculated as the sum of the squares of the deviation of each observation from the mean, divided by $n - 1$ (where n is the number of observations). The total variance is known as the phenotypic variance, part of which is accounted for by genotypic variance. In genetical research, the exact proportion of the variance that can be accounted for by genotypic variance can sometimes be ascertained (but by no means always – *see* nature versus nurture). *See also* covariance; standard deviation.

Variation, coefficient of, a measurement of the amount of variation within a population. It equals the standard deviation expressed as a percentage of the value of the mean.

Variegation, patterns, regular or irregular, of different colouring on the leaves of plants. Environmental causes include viral infection and mineral deficiency; genetic causes include alleles controlling regular patterns (often of red or yellow on green, e.g. in *Coleus*), a >chimaera genetical constitution, or local gene-switching by >jumping genes.

Vavilov, Nikolai Ivanovich, (1887–*c*. 1943), Soviet geneticist. Vavilov learned his genetics from >Bateson in England, and on his return to the Soviet Union, he was soon promoted to director of the Lenin All-Union Academy of Agricultural Sciences. He researched the origins of cultivated plants, which he believed could be identified from the present-day distribution of wild species. He travelled all over the world collecting live specimens, and eventually had a quarter of a million separate varieties in cultivation. In his book *The Origin, Variation, Immunity and Breeding of Domestic Plants* (English edition 1951), he stated that the earliest domestication of plants had occurred in two places, Ethiopia and Afghanistan. His research also helped to produce improved strains of wheat without which the famines in the Soviet Union would have been even worse. A convinced Mendelian, Vavilov fell foul of >Lysenko and the official doctrine of the inheritance of acquired characteristics. In 1940 he was sacked and sentenced to death; the sentence was commuted to imprisonment in a labour camp in Siberia, where he is believed to have died of starvation in 1943.

Vector. There are two meanings. First, in >genetic engineering it is the means whereby a gene or DNA sequence is carried into the host cell. Vectors, which must be capable of being replicated by the host cell on arrival, are usually either bacterial >plasmids or viral DNA molecules.

Althernatively, a vector is any organism that is responsible for transferring a pathogen (disease-causing agent) or parasite from one host to another. For example, the mosquito that carries the *Plasmodium* parasite causing malaria is a vector.

Vegetative propagation, reproduction without a sexual process. The simplest form is *equal cell division*, as in bacteria; after this comes *unequal budding*, with a small new cell growing after it has budded off the parent cell (as in yeast). In higher plants, vegetative propagation occurs naturally as multiplication of bulbs, corms, and runners, and artificially as grafts. The products are always a >clone of the original individual.

Venoms, the poisons produced by snakes, spiders, etc. In snakes, the venom consists of a mixture of poisonous proteins which are either haemolytic (i.e they break down the blood cells and/or prevent clotting) or neurotoxic (i.e they attack the nerves, particularly those that control breathing). The actual proteins and the balance of haemolytic and neurotoxic effects of smoke venoms vary almost from species to species, so that anti-venoms have to be specific to the snake in question. The genetics of the various venoms is not precisely known.

Vertebrates, all animals with backbones. These include fish, birds, reptiles, amphibians and mammals. *See* classification.

Vertical transmission, the usual type of genetic inheritance, from one generation to the next. It is contrasted with >horizontal transmission.

Vestigial organ, any organ that persists in a later evolutionary form after its original purpose has disappeared ('organ' here has the meaning of any anatomically distinct structure). Examples are the dew-claws in dogs and the coccyx ('tail') in humans. Vestigial organs were taken by >Buffon, >Lamarck and others as evidence of the evolution of species, as indeed they are.

Virulence, the ability of a pathogen (a virus, bacterium or other micro-organism) to produce a disease. It depends partly on its capacity to invade the host cell and multiply, and partly on its ability to produce toxins. Virulence tends to be favoured by natural selection, as virulent strains are often the most successful at reproducing themselves; but beyond a certain point virulence is a disadvantage to a strain of pathogens, as they kill off their hosts too soon for their own good.

Viruses, the simplest of living things, consisting of a core of either DNA or RNA inside a protein coat. They are not classified as truly living organisms, as they are not capable of self-reproduction but rely on the reproductive capability of their host cell (which may be a bacterium or a cell within a plant or animal). A virus enters a cell by attaching its protein capsule to the outside of the cell wall and injecting just its DNA or RNA into the cell. Then either the virus multiplies rapidly within the cytoplasm until the cell bursts, releasing large numbers of viruses to attack other cells (this is known as >lysis), or the virus integrates its own genes into the host cell's genome and is replicated while remaining harmless itself (this is known as >lysogeny). In this latter state the virus is known as a >provirus. A virus typically has a very small number of genes, perhaps as few as four, two to code for the protein coat and two to help the virus insert its own DNA into the host's genome.

Those viruses that infect animals include:

- *DNA genomes*: >adenoviruses, >herpesviruses, >poxviruses, >parvoviruses.
- *RNA genomes*: reoviruses, retroviruses.

There are also many types of plant virus. The viruses that attack bacteria are called >bacteriophages.

Vitalism, the theory that living matter is intrinsically different from non-living matter. The French philosopher Henri Bergson proposed (in

L'Evolution créatrice, 1907) that the *élan vital* ('vital spirit') in all living things is the driving force behind >evolution. This idea was taken up by George Bernard Shaw. Similarly the German biologist Hans Driesch taught that the development of the embryo is driven by some vital element in living matter. Vitalism had a tenuous existence in scientific circles in the first half of the twentieth century, because the properties claimed for whatever was the molecule of heredity seemed unlikely to be possessed by any known chemical compound. The discovery of the structure of >DNA must have finally disposed of vitalism.

Vitamins, a group of substances that are necessary for metabolism but only in minute quantities. They are chemically very diverse, but are divided into two groups: (*a*) water-soluble vitamins – the Bs and C; (*b*) fat-soluble vitamins – A, D, E and K. Vitamins work as co-enzymes within the >enzyme system, enabling the various essential chemical reactions within cells to take place, e.g. the production of energy, the synthesis of proteins (for tissue-building, hormones, etc). There are some genetically caused vitamin deficiency diseases, e.g. the form of rickets that results from an inability to utilize vitamin D.

Viviparous, bringing forth live young. This is the latest phase in the evolution of reproductive methods. *See also* oviparous; ovoviviparous.

Vries, Hugo de (1848–1935), Dutch botanist; one of the three rediscoverers of >Mendel's laws. De Vries was professor of botany at Amsterdam for 40 years, and was the first person to study evolution by experimental methods. He introduced the word 'mutation', but in a different sense to its modern one; he believed that evolution can proceed either gradually by natural selection or in sudden jumps ('mutations'). De Vries's experiments led him to discover the same 3:1 ratio in F_2 hybrids that Mendel had done, and he came up with the same theory to explain it. Searching the literature, he found Mendel's paper published in 1866; he then discovered that >Correns and >Tschermak had come to the same idea and had also found Mendel's paper. In 1900, de Vries arranged for Mendel's paper to be republished and publicized.

W

W chromosome, one of the sex chromosomes in birds and some insects, in which the female is the heterogametic sex (i.e. has two different sex chromosomes). The female has chromosomes WZ, and the presence of the W causes femaleness to develop (*see* sex determination); the male is ZZ.

Wallace, Alfred Russel (1823–1913), British naturalist; co-discoverer of the principle of evolution through natural selection. Wallace was self-educated and was working as a surveyor when he read Darwin's notes of the *Beagle* voyage, which inspired him to become a naturalist. He went with Henry >Bates to South America, but unfortunately he lost his collection of live specimens when his ship caught fire on the way home. Then he went to the Malay Archipelago, and it was there that he had the idea about natural selection. He had arrived at it by exactly the same mental route as Darwin: during his travels, he had observed the variability of species, and then he had read Malthus. He sent his essay to Darwin in 1858, and Darwin's friends arranged for the two accounts to be presented to the public simultaneously (see p. 17). Wallace was never resentful that the theory became known as Darwinism. He differed from Darwin in doubting whether natural selection applied to human beings (this was perhaps because Wallace was a spiritualist). Wallace systematized biogeography (an important area of evidence for evolution) in *The Geographical Distribution of Animals* (1876) and *Island Life* (1880), and proposed the 'Wallace line', the imaginary line that divides the faunas of Australasia and South-east Asia.

Waltzing mice, those that have an inherited defect in the balancing mechanism in the inner ear, causing them to turn in small circles. There are several mutant genes that give this effect. Waltzing mice were known in China as early as 80 BC, in the Han dynasty.

Warfarin, an anticoagulant drug. Its main use is in treating arterial disease, but it is also well known as a poison used against rodents. In the 1950s rats in Britain evolved resistance to warfarin, and the immune rats spread very rapidly (the rapidity of the spread being related to the severity of the selection pressure, i.e. the amount of poisoning being carried out). They became known as 'Super Rats', an alarmist name for animals that were perfectly normal except for one biochemical trait. There should have been more alarm at the danger of substantial risk to public health that comes when evolution hits back at human attempts to wipe out pests. *See* resistance.

Watson, James Dewey (1928–), American microbiologist; director of the Cold Spring Harbor Laboratory, Long Island, New York, since 1968 and, more recently, one of the prime movers of the Human Genome Project (*see* pp. 36–8), from which he resigned as director in 1992; co-discoverer with Francis >Crick of the structure of >DNA. Watson studied at the University of Chicago and then researched bacteriophage genetics at the University of Indiana, but at a very early stage he became obsessed with the idea of solving what he saw as the fundamental problem in biology: what *is* a gene? He went to Cambridge University in 1951 to do a second doctorate, and joined the Medical Research Council's unit at the Cavendish Laboratory. He and Crick had some clues to follow in their quest for the gene, including >Avery's discovery that DNA is the genetic material, >Chargaff's rules on the ratios of the bases in DNA, and >Franklin's X-ray diffraction photographs, but their breakthrough only came when they tried building a scale model of the DNA molecule out of pieces of cardboard and wire. He and Crick shared the 1962 Nobel prize for physiology or medicine with Maurice Wilkins (a colleague of Franklin's). Watson tells the story of the discovery of the DNA structure in his entertaining book *The Double Helix* (1968), which is however, ungenerous to several people, including Crick and Franklin. His scientific writings include the important *Molecular Biology of the Gene*, now in its fourth edition (1986). *See also* pp. 29–30.

Weismann, August (1834–1914), German biologist. Weismann trained as a doctor but became professor of zoology at Freiburg. He was a brilliant embryologist, but had to give up his researches on the development of insects when his eyesight deteriorated. Turning to theoretical biology, he proposed that heredity works by means of 'germ plasm' which is unchangeable and is not formed anew but is passed from one generation to the next. This 'germ plasma', according to Weismann, controls all aspects of the development of the organism (*see* germ plasm theory); he was consequently a firm opponent of the theory of the inheritance of acquired characteristics. Weismann saw a difficulty in that, if germ plasm were inherited equally from each parent, the amount of it would double in every generation unless there were some mechanism that reduced it. He was vindicated when >Hertwig observed meiosis, which provides exactly the appropriate reduction mechanism and locates the 'germ plasm' in the chromosomes. Weismann then proposed that one source of biological variation is the different combination of chromosomes at meiosis. His ideas were far ahead of their time, and in perfect accord with Mendel's work (then published but unknown).

Wheat. *See Triticum aestivum.*

Whole-arm fusion. *See* Robertsonian translocation.

Wild type, the allele that is deemed to be the one that occurs in the species in its original state, as opposed to subsequent mutations, either spontaneous or induced. The wild-type allele may be either dominant or recessive in relation to a mutant allele. Its symbol is +.

W

Wilson, Edward Osborne (1929–), American entomologist; has taught at Harvard University since 1953; father of >sociobiology. Wilson was already one of the world's leading authorities on social insects when he published *Sociobiology: The New Synthesis* (1975). He was unshaken by the acrimonious controversy that followed, and has continued to develop the same ideas about the importance of genetics in determining social behaviour.

Wobble, the fact that a single >anticodon of tRNA can recognize different codons in DNA, provided that the difference is in the third base. For example, the anticodon for cystine, ACG, can recognize and bind itself to either UGC or UGU (*see* genetic code). Wobble was first recognized by Francis >Crick.

Wright, Sewall (1889–1988), American geneticist. Wright became interested in genetics via agriculture, and his early work was on the best ways to use inbreeding and cross-breeding to improve animal stocks. Turning to the study of populations of animals in the wild, he discovered that if a population is very small, genes can be lost for ever from the gene pool, not because they were inferior but because of mere chance. This is now known as the >Sewall Wright effect.

X

X chromosome, one of the sex chromosomes. In humans and many other creatures, the female has two X chromosomes and the male an X and a Y (*see* sex determination). The human X chromosome is the seventh largest in the human karyotype, nearly three times the size of the Y chromosome. It carries a large number of genes, which are said to be X-linked (*see* X linkage). *See also* Lyon hypothesis; Ohno's rule.

X linkage, the situation in which a gene is located on the X chromosome. In males, with sex chromosomes XY, any allele on the X chromosome behaves as a dominant, because there is no corresponding allele to mask it on the other chromosome (the Y does not behave as homologous to the X). A male passes on to all his daughters any gene that he carries on the X chromosome, but his sons will not inherit it. A female who carries the same allele on one of her X chromosomes will not be affected, if the gene is recessive as most X-linked genes are. Such a female is known as a carrier, and passes on the gene to half of her offspring, sons and daughters alike: males who receive the gene will inherit the condition, females will be carriers (*see diagram*).

It is possible for a female to inherit an X-linked gene simultaneously from her father (he will have the trait in question) and from her carrier mother, but this is rare (how rare depends on the frequency of the X-linked allele in the population). The ginger colour in cats is X-linked, and most ginger cats are male, but the occasional female ginger cat can be found (*see also* tortoiseshell cat).

Over 100 genes are known to be X-linked in humans, including >colour blindness, >Duchenne muscular dystrophy and >haemophilia. Occasionally an X-linked gene is dominant; its pattern of transmission is similar,

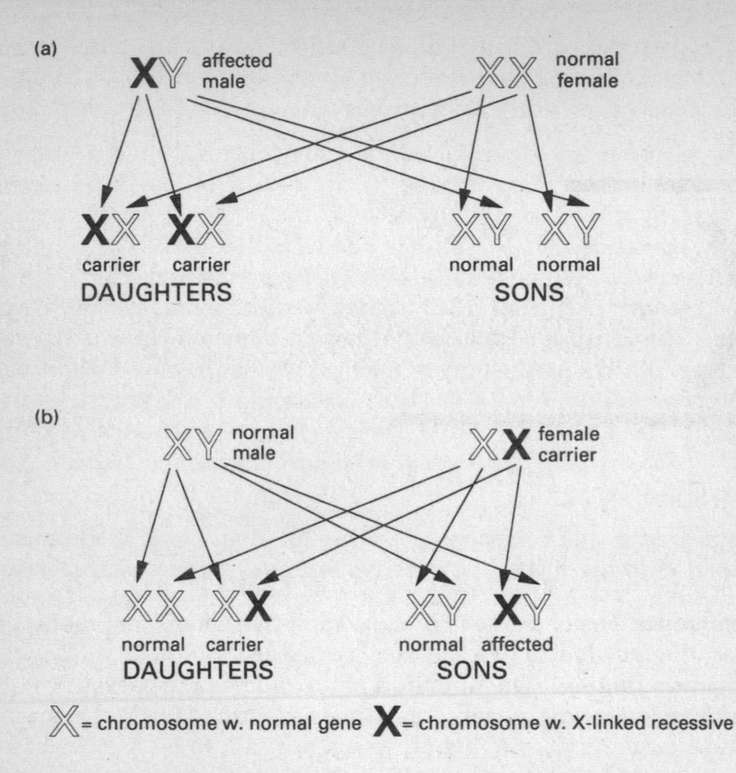

in that a father cannot pass it on to his sons. An X-linked trait is not to be confused with a >sex-limited trait. *See also* Barr body; Lyon hypothesis; Ohno's rule; sex chromosomes.

Xenopus laevis, the African clawed toad. It has been used for experiments in developmental genetics, because its tadpole (which is in effect the embryo) is robust and easy to keep. It is also useful in experiments which involve the removal of the cell nucleus, as *Xenopus*'s nucleus is large; these experiments have included the earliest successful clones. There are now a number of inbred strains for genetic use. *Xenopus* has a diploid chromosome number of 36.

Xeroderma pigmentosum, a human disease in which the skin is excessively vulnerable to bright sunlight, resulting in a high incidence of skin cancers. The underlying cause is an inherited inability to repair the damage that ultraviolet rays do to the DNA in the skin cells. The disorder is inherited as an autosomal recessive, and is one of the few clear-cut examples of a genetic basis for cancer.

X-ray crystallography, the technique used for investigating the three-dimensional structure of large molecules such as proteins and nucleic acids. The wavelength of the X-rays used is similar to the inter-atomic distances in these molecules, and the position of atoms can be deduced from the patterns of diffraction seen when X-rays are passed through them. X-ray diffraction pictures taken by Rosalind >Franklin contributed to Crick and Watson's discovery of the structure of >DNA (*see* p. 28).

X-rays, electromagnetic radiation at frequencies between ultraviolet light and gamma-rays, about 10^{-8} to 10^{-11} m wavelength. X-rays were discovered by the German physicist Wilhelm Roentgen in 1895. The American geneticist H. J. >Muller discovered in 1926 that they were a potent means of inducing >mutations in *Drosophila* and other organisms, and he became very concerned that the widespread use of X-rays in medicine would cause serious mutations in humans. He was heeded; no longer were X-rays used on every medical pretext however slight or even as a commercial gimmick (e.g. there used to be X-ray machines in shoe shops for fitting children's shoes). Nowadays X-rays are used as sparingly as possible, particularly in the region of the testicles and ovaries. *See also* X-ray crystallography.

XYY syndrome, the condition of having an X and two Y chromosomes. Boys born with this develop physically as normal males and are fertile, but are taller than average (by about 6 in/15 cm). However, there is evidence that XYY men are five times more likely than XYs to be mentally subnormal. What is more, they are more likely to start displaying criminal behaviour as boys. According to findings in the United States, XYY boys tend to commit crimes of impulsive aggression (especially those leading to criminal damage rather than attacks on people).

When this research was first published, there was much talk about XYY being a 'criminal karyotype', and in the 1960s there were notorious murder trials in four different countries at which the defendants pleaded that they were not criminally responsible because of their XYY karyotype. In the French case, the man was convicted but given only a light sentence owing to that mitigating circumstance; in the English and German cases, the plea was ruled out of order and the men convicted; in the Australian case, the man was found not guilty by reason of his extra Y. It must be emphasized that the American research showed only statistical trends among the whole XYY population in the US; the great majority of XYY people (about 90 per cent) have no behavioural or mental difficulties at all.

Y

Y chromomsome, one of the sex chromosomes. In humans and many other creatures, the male has an X and a Y, while the female has two XXs; it is the presence of the Y that causes maleness to develop (*see* sex determination). The gene responsible for this acts by accelerating the development of the previously undifferentiated urinogenital tract of the embryo to form first the testes and then the male ducts and penis: it is an example of a >genetic switch. The sex-determining gene is on the short arm of the Y chromosome; individuals who are XY but have lost the short arm of the Y develop as female. The longer arm of Y can apparently be lost with no effect on the phenotype, so it is believed that there are no general-purpose genes on it. The Y chromosome is the third smallest in the human karyotype.

Yeast. *See Saccharomyces cerevisiae.*

Z

Z chromosome, one of sex chromosomes in birds and some insects, in which the male is the homogametic sex (i.e. has two identical sex chromosomes). Males are ZZ and females are WZ. *See* sex determination; W chromosome.

Z DNA, DNA in which the double helix is wound left-handed rather than right-handed as normal. The sugars do not fit together so well that way round, so the molecule has a jagged appearance. Z DNA has been studied in *Drosophila* chromosomes, and it has been suggested that it has a role in gene regulation.

Zea mays, the staple crop known in American English as 'corn' and in British English as 'maize' or 'sweetcorn'. It was the subject of early genetical research because of the economic importance of producing improved strains. It also happens to be a convenient plant to study in that large numbers can be grown on small plots of land, and the variants in the kernels are easy to observe and measure. Some important basic genetic discoveries have been made in *Zea*, including jumping genes.

Zygote, the individual as it exists at the moment of fertilization, i.e. the diploid cell that results from the fusion of two gametes. It can also be a theoretical entity: one could calculate the number of possible different zygotes that could result from a given mating (*see* hybrid).

Zygotene, one of the stages in >meiosis (one of the two processes of cell division). *See* diagram at meiosis.

Recruitment

396 88 56 ⟶ check status of application

57